Studies in Systems, Decision and Control

Volume 280

Series Editor

Janusz Kacprzyk, Systems Research Institute, Polish Academy of Sciences, Warsaw, Poland

The series "Studies in Systems, Decision and Control" (SSDC) covers both new developments and advances, as well as the state of the art, in the various areas of broadly perceived systems, decision making and control–quickly, up to date and with a high quality. The intent is to cover the theory, applications, and perspectives on the state of the art and future developments relevant to systems, decision making, control, complex processes and related areas, as embedded in the fields of engineering, computer science, physics, economics, social and life sciences, as well as the paradigms and methodologies behind them. The series contains monographs, textbooks, lecture notes and edited volumes in systems, decision making and control spanning the areas of Cyber-Physical Systems, Autonomous Systems, Sensor Networks, Control Systems, Energy Systems, Automotive Systems, Biological Systems, Vehicular Networking and Connected Vehicles, Aerospace Systems, Automation, Manufacturing, Smart Grids, Nonlinear Systems, Power Systems, Robotics, Social Systems, Economic Systems and other. Of particular value to both the contributors and the readership are the short publication timeframe and the world-wide distribution and exposure which enable both a wide and rapid dissemination of research output.

** Indexing: The books of this series are submitted to ISI, SCOPUS, DBLP, Ulrichs, MathSciNet, Current Mathematical Publications, Mathematical Reviews, Zentralblatt Math: MetaPress and Springerlink.

More information about this series at http://www.springer.com/series/13304

Jorge E. Hernández · Janusz Kacprzyk
Editors

Agriculture Value Chain—Challenges and Trends in Academia and Industry

RUC-APS Volume 1

 Springer

Editors
Jorge E. Hernández
University of Liverpool
Management School
Liverpool, UK

Janusz Kacprzyk
Systems Research Institute
Polish Academy of Sciences
Warsaw, Poland

ISSN 2198-4182 ISSN 2198-4190 (electronic)
Studies in Systems, Decision and Control
ISBN 978-3-030-51049-7 ISBN 978-3-030-51047-3 (eBook)
https://doi.org/10.1007/978-3-030-51047-3

This Springer imprint is published by the registered company Springer Nature Switzerland AG
The registered company address is: Gewerbestrasse 11, 6330 Cham, Switzerland

Preface

This volume is the first collection of works which have been done within RUC-APS Project "Enhancing and Implementing Knowledge Based ICT Solutions Within High Risk and Uncertain Conditions for Agriculture Production Systems", run through via the H2020 RISE-2015 Call over the period from October 3, 2016 to October 2, 2020. The project has been led and coordinated by Dr. Jorge E. Hernández from the University of Liverpool, UK, and the consortium comprised 13 participants from five EU countries (France, Italy, Poland, Spain, and UK), and five partners from two-thirds countries (Argentina and Chile), which in total involves more than 150 researchers.

The project has been proposed in response to an increasing necessity to cope in agricultural systems, notably agriculture values chains, with abrupt changes in resource quality, quantity and availability, especially during unexpected environmental circumstances, such as uncertain weather, pests and diseases, volatile market conditions and highly varying commodity prices. The solution of these problems, often just mitigation, needs a comprehensive approach considering risks throughout the whole food agricultural lifecycle value chain to attain resilience and sustainability.

Farmers, crop growers and animal breeders should be supported by tools and techniques, possibly implemented as computer systems, notably computer decision support systems (DSS), which would help them to manage their risks and to cope with the uncertain availability of information to arrive at solutions which could be considered to be good from the economic, technological, social, etc., points of view. This implies a need for an innovative technology-based knowledge management system for dealing with the relevant agricultural information, that is, with collecting, storing, processing, and disseminating information about uncertain environmental conditions that affect agricultural decision making.

RUC-APS has been meant to develop, in a collaboration between the academic and commercial teams, an effective and efficient system which would make it possible to discover knowledge of the full agricultural life cycle-based decision making process, both from a more general point of view by treating this process as a whole, and in a more detailed way to attain the key impact results at every stage of the agriculture related processes.

Therefore, the idea of RUC-APS has implied the launching and development of a high impact research project in order to integrate real life-based agriculture requirements, alternative land management scenarios, unexpected weather and environmental conditions, volatility of markets for agricultural products, as well as supporting innovation in the development of agriculture production systems, operations, logistics, and supply chain management, all this viewed from the points of view of both the end users, intermediaries, and end customers. This has been implemented through the integration of standard and customized solutions for facilitating the cooperation and collaboration over the agriculture value chain.

As we have already mentioned, uncertainty and risk are two crucial issues for agriculture, and their importance is growing in view of more and more difficult and volatile conditions to be faced by the agricultural systems and value chains. Traditionally, since the 1970s agricultural economics has primarily focused on seven main topics:

- agricultural environment and resources;
- risk and uncertainty;
- food and consumer economics;
- prices and incomes;
- market structures;
- trade and development; and
- technological change and human capital.

These topics still remain as important areas of interest and research. Uncertainties concerning all the above topics are usually a cause for ineffective and inefficient decision-making processes for farmers and all participants in the agriculture value chain. However, recently, a new kind of uncertainties, notably implied by climate changes, has become more and more pronounced and yet has not been adequately addressed in the agriculture and decision-making literature in spite of the fact that will certainly has a tremendous impact on all ecosystems.

However, the analysis of these uncertainties is difficult because there is a lack of complete knowledge or historical parallels in terms of the uncertainty about the reactions of ecosystems to climate changes. Moreover, more fundamentally, there is considerable uncertainty as to how high a degree of climate change we face. One the main goals for agriculture and the whole food sector is to find right responses and strategies to the challenge of a much faster rate of climate change we face, for instance by reducing global emissions by planting trees, reducing tillage, increasing soil cover, improving grassland management, using fertilizer more efficiently, etc. to just mention a few.

By extending the current state-of-the-art knowledge in the field, RUC-APS aims at providing new solutions for the agriculture-based decision making through the development of tools and techniques, implemented possibly in modern decision support system settings, to integrate in the analysis and solution process real-life agriculture-based value chain requirements, land management alternatives for a variety of scales, unexpected weather and environmental conditions as well as the innovation for the development of agriculture production systems and their impact over the end users.

To be more specific, the activities of RUC-APS have concentrated on the following more detailed topics which have practically covered all crucial elements for the satisfaction of the project's goals, that is:

- To support the varietal behaviour in a variety of agro-climatic zones in terms of cost and production cycle,
- To optimize the agronomic management by understanding the incorporation of a specialized machinery in the processes of sowing, transplanting and fertilizing,
- To analyze and refine quality standards to ensure the safety of the final agriculture-based product under high risk and uncertain conditions,
- To evaluate the environmental conditions for enhancing soil management technologies and planting management,
- To model and optimize smart and innovative collaborative production/transport planning solutions of horticulture products across the full value chain,
- To generate horizontal and vertical knowledge exchange mechanisms across the agriculture value chain, and the implementation of these ambitious and comprehensive goals has been proposed to be through very efficient exchange programs between the teams meant as:
- To draw together outstanding academic research teams from Europe and South America, with the benefit of achieving synergy from their diverse multidisciplinary skills and expertise in the field of agronomy, agriculture, uncertainty, risk, supply chain and logistics,
- To build joint collaborative projects for the development of sustainable international agriculture management under high risk and understanding with high efficiency, resilience and integration in terms of risk and cost that are subjected to various unpredictable environments in a variety of latitudes,
- To create a platform for research training and transfer of knowledge activities, both for the participating teams and open to the general public, crossing inter-sector boundaries to ensure that there is a sufficient number of people trained in this field to handle the existing and emerging challenges, and to ensure the EUs leading position in the world in combination with a comprehensive and diversified expertise,
- To disseminate knowledge to the larger international communities to maximize the impact of this network, and ensure the academic, industrial and governmental relevance of its research activities,
- To provide researchers, in particular at an early stage in their careers, with an intensive training through participation programs as well as complementary skill training, in order to become in the future trainers in a multidisciplinary field of academic, industrial and societal importance, and by so doing improving the career prospects of young researchers,

- To establish collaborative mechanisms for long-term partnerships between European and South American researchers and institutes on the resilience and sustainability agriculture management by integrated information communication technologies-based systems in order to support the value chain participants in their decision-making process under high risk and uncertain conditions, where small and medium farmers will be able to interact better with issues exemplified by climate change, unpredictable environment and economic conditions, increasing economic upheaval and other challenges for the twenty-first century in the agribusiness and innovation approaches.

This volume is planned as the first volume being a result of a very successful collaboration of researchers and practitioners from many countries facing different challenges in their agriculture systems that have proceeded within RUC-APS. It contains seven papers which constitute the first presentation of results in the areas of decision making, decision processes, and support systems, notably using new Web and semantic technologies, various aspects of uncertainty management and risk management and mitigation, and analyses of specific cases of crop production.

The contents of the consecutive contributions can be briefly summarized as follows.

Matías Urbieta, Sergio Firmenich, Pascale Zaraté, and Alejandro Fernandez ("Web-Augmentation, Design Thinking, and Collaboration Engineering to Foster Innovation in DSS for Agriculture: A Case Study") are concerned with decision making in agriculture in modern, nontrivial settings in which the reliance on software, for example, to gather important information or to weigh alternatives, is crucial. Innovations in information systems in agriculture is a challenging area of very active research efforts. Existing software products, frequently implemented as Web applications, are found to lack functionalities, for example to support collaboration. The augmentation of the Web is a widely adopted technique for enhancing existing applications with new features which are not available out of the shelf. Design thinking has proven to be an effective and efficient tool to support innovation in many domains. Collaboration engineering is an effective and efficient means to reuse design experience of collaboration strategies. This work presents an approach on how to involve the end users in the process of enhancing the exiting Web software to yield incremental innovations. The approach relies on design thinking, Web augmentation and collaboration engineering. The approach has been tested on a case study letting the end users add a collaboration support to the system, a new feature which has not been previously provided.

M. M. E. Alemany, Ana Esteso, A. Ortiz, J. E. Hernández, A. Fernández, A. Garrido, J. Martín, S. Liu, G. Zhao, C. Guyon, and R. Iannacone ("A Conceptual Framework For Crop-Based Agri-food Supply Chain Characterization Under Uncertainty") deal with the Crop-based Agri-food Supply Chains (AFSCs) which are complex systems that face multiple sources of uncertainties which can cause a significant imbalance between the supply and demand in terms of product varieties, quantities, qualities, customer requirements, times and prices, etc. all of which greatly complicate the management of the AFSCs. Poor management, or mitigation, of these sources of uncertainty in such AFSCs can have a negative impact on quality, safety, and sustainability by reducing the logistic efficiency and increasing the waste. Therefore, it becomes crucial to develop models in order to deal with the key sources of uncertainty. For this purpose, it is necessary to precisely understand and define the problem considered. The characterization process of this domains is also a difficult and time-consuming task when the right directions and standards are not available. The authors propose a conceptual framework which combines those aspects that are relevant for an adequate management under uncertainty of the crop-based agri-food supply chains (AFSCs).

L. Antonelli, G. Camilleri, C. Challiol, A. Fernandez, M. Hozikian, R. Giandini, J. Grigera, A. B. Lliteras, J. Martin, D. Torres, and P. Zarate ("An Extension to Scenarios to Deal with Business Cases for the Decision-Making Processes in the Agribusiness Domain") consider an important issue of how to push innovation through information and communication technology in the agri-business field being aware of the fact that working closely with farmers (experts) is essential. Of a particular importance is to be able to systematically capture the knowledge of the farmers in order to analyze, propose, and design innovation artefacts (in terms of software applications). The authors use scenarios to capture the knowledge of the experts that is elicited in early meetings prior to the definition of requirements. At those early stages, there are many uncertainties, and the use of decision support is well justified. Thus, an extension of the scenarios for dealing with uncertainties is proposed. The scenarios are described in natural language, and it is very important to have an unbiased vocabulary. The authors propose to complement the scenarios with a specific glossary, the Language Extended Lexicon, which is also extended to the field of decision support. According to the model's life cycle, every stage has a testing-related stage. Thus, a set of rules is also proposed to derive tests from the scenarios. Summing up, the authors propose: an extension to the scenarios and the Language Extended Lexicon templates, a set of rules to derive tests, and an application to support the proposed technique. The solution of some case studies confirms that the results are promising.

C. Jana, G. Saavedra, N. Calabrese, J. P. Martinez, S. Vargas, and V. Muena ("Functional Value in Breeding Integrated to the Vegetables Value Chain as Part of Decision Making") are concerned with recent trends of a great demand for healthy food containing added functional value implied by the fact that people have begun to change their diet with increments in the consumption of foods rich in antioxidants compounds, vitamins, minerals, etc. Since vegetables are one of the most important sources of bioactive compounds, then the development of new varieties with an enhanced content in bioactive compounds has become an increasingly important breeding objective. Breeding programs should be focused, not just on pest resistance or tolerance, yield, etc., but on the added value of the nutraceutical compounds which are particularly beneficious for the human health. There have been efforts in some vegetable categories to generate new breeds with an increased content of bioactive compounds. The antioxidants are an added value for the consumer, some of them are already commercial, but in general there is still a long way to go. The authors analyze goals and advances in vegetable breeding, conventional and with use of biotechnology, and their integration into the value chain, specifically with respect to three important crops: artichoke, tomato and lettuce.

G. Saavedra, S. Elgueta, E. Kehr, and C. Jana ("Climate Change Mitigation Using Breeding as a Tool in the Vegetables Value Chain") consider important issues related to the fact that agriculture is a climate dependent activity so that it is affected by both climate change and also contributes to climate change. So, agriculture is both a victim (because of negative effects of the global warming on food supply) and a perpetrator (one of the main causes of climatic change). To producing more food for a starving world, the agriculture requires high amounts of nitrogen-based fertilizers, which in turn releases nitrous oxide (N_2O) emissions, one of the greenhouse gases. The carbon dioxide, an important Greenhouse gas, for instance, can be beneficious for some vegetable species. There are other impacts of climate change in the vegetable production, not only in plant physiology. The quality of plants and fruits is affected by elevated temperatures which affects pollination and fruit set. There are other impacts on vegetable production systems, for instance, extended growing seasons because of warmer springs and autumns. New areas of vegetable production, where has been so far impossible to produce larger amounts of quality vegetable because of too low temperatures, become feasible. Unfortunately, the mitigation of climate change is not an easy task. It is concerned with the human conscience too. Breeders may do huge efforts to generate varieties tolerant to many stresses. A challenge is to generate varieties tolerant to high-temperature stress and low input production, with a high rusticity and being able to respond to biotic and abiotic stresses caused by climate change.

S. Elgueta, A. Correa, M. Valenzuela, J. E. Hernández, S. Liu, H. Lu, G. Saavedra, and E. Kehr ("Pesticide Residues in Vegetable Products and Consumer's Risk in the Agri-food Value Chain") address issues related to a large increase of demand for food which occurs worldwide due to the huge population growth and better living standards. The use of pesticides in modern agriculture is necessary for most crops to guarantee an adequate production but an improper and excessive use of pesticides in agriculture implies an increasing health risk. Some products can be a source of pesticide residues, and their ingestion represents a potential source of diseases and negative effects to the human body and hence human health. The authors present principles and methods for health risk assessment of pesticide residues in vegetables and fruits. The exposure assessment of pesticide residues is an essential method to quantify the chemical risks. In addition, the dietary exposure assessment combines the food consumption data with the pesticide concentrations in the vegetables and fruits. The aim of the consideration is to support the decision-making process in the agri-food value chain to improve the food safety enforcement and reduce impacts on the human health. The assessment can be developed for the problem formulation and the interactions between the risk assessor, risk manager, stakeholders, and decision makers. In addition, the assessment based on a scientific evidence facilitates the decision-making process in order to determine the impacts of pesticide residues on human health and set up their maximum residue limits.

J. P. Martinez, C. Jana, V. Muena, E. Salazar, J. J. Rico, N. Calabrese, J. E. Hernández, S. Lutts, and R. Fuentes ("The Recovery of the *Old* Limachino Tomato: History, Findings, Lessons, Challenges and Perspectives") present a summary of some aspects related to the recovery of the *Old* Limachino Tomato, a special tomato cultivated in the Limache Basin located in the Valparaíso Region, Chile. This specie disappeared completely from the market some 45 years ago, and the paper presents a very interesting account of its recovery and reintroduction process. The authors highlight both technical and commercial aspects of this process that has given back to the Chileans a tasty, healthy fruit with high functional qualities. Moreover, it has been shown that issues related to the protection of the Peasant Family Farming of the cultivation area and the improvement of post-harvest quality have been crucial for the sustainability of the business. In particular, it has been claimed that from a sustainable agriculture development perspective, there is an urgent need to increase the tolerance of this species to salinity stress in order to increase the productivity and reduce the use of water and fertilizers. In addition, the use of trademarks and protection certificates of the tomatoes is of relevance.

We hope that the inspiring and interesting contributions, included in this volume, will be of much interest and use for a wide research community, not only in agriculture and food industry.

We wish to thank the contributors for their great works. Special thanks are due to anonymous referees whose deep and constructive remarks and suggestions have helped to greatly improve the quality and clarity of the contributions.

And last but not least, we wish to thank Dr. Tom Ditzinger, Dr. Leontina di Cecco, and Mr. Holger Schaepe for their dedication and help to implement and finish this important publication project on time, while maintaining the highest publication standards, as well as to Miss. Rachel Brockley, for her outstanding support and contribution to manage the RUC-APS project.

The collection of the chapters provides in-depth insights into challenges that the contemporary agri-food value chain sector facing in terms key sources of uncertainties, such as climate change. But also addresses how multi-sector cross-knowledge and multidisciplinary collaborations can be translated into tangible research outcomes. Therefore, the selected studies present a full spectrum of the current international agri-food value chain research agenda, from strategies of innovation in agri-food supply chains, to information communication and technology-based decision support systems to performing sustainable agri-food operations. We expect that this book edition will support both academics and practitioners to developing sustainable supply chain research and practice, considering both theory and practice.

Liverpool, UK   Jorge E. Hernández
Warsaw, Poland   Janusz Kacprzyk
Winter 2019   Editors

Acknowledgement

RISK-UNCERTAINTY-COLLABORATION
AGRICULTURE PRODUCTION SYSTEMS

Partially supported by the H2020 RISE-2015 project:

RUC-APS: Enhancing and Implementing Knowledge Based ICT Solutions Within High Risk and Uncertain Conditions for Agriculture Production Systems

Contents

Web-Augmentation, Design Thinking, and Collaboration Engineering to Foster Innovation in DSS for Agriculture: A Case Study

Matías Urbieta, Sergio Firmenich, Pascale Zaraté, and Alejandro Fernandez

Abstract Decision making in agriculture increasingly relies on software, for example to gather important information or to weight alternatives. Information systems innovation in agriculture is a challenging and very active area. Existing software products, frequently implemented as web applications, are found to lack functionality, for example to support collaboration. Augmenting the web is a widely adopted technique for enhancing existing applications with new features which are not available out-of-the-shelf. Design thinking has proven to be an effective tool to support innovation on many domains. Collaboration Engineering is an effective means to reuse design experience of collaboration strategies. This work presents an approach to involve end-users in enhancing exiting web software to produce incremental innovations. The approach relies on Design Thinking, Web Augmentation and Collaboration Engineering. The approach was successfully tried in a case study letting end-users add collaboration support to a system that did not provide it.

Keywords Agriculture · Software · Augmentation · Collaboration

M. Urbieta (✉) · S. Firmenich · A. Fernandez
LIFIA, Facultad de Informática, Universidad Nacional de La Plata, La Plata,
Argentina
e-mail: matias.urbieta@lifia.info.unlp.edu.ar

S. Firmenich
e-mail: sergio.firmenich@lifia.info.unlp.edu.ar

A. Fernandez
e-mail: alejandro.fernandez@lifia.info.unlp.edu.ar

M. Urbieta · S. Firmenich
CONICET, Buenos Aires, Argentina

A. Fernandez
CICPBA, La Plata, Argentina

P. Zaraté
Toulouse University, IRIT, 2 Rue du Doyen Gabriel Marty, 31043 Toulouse Cedex 9,
France
e-mail: pascale.zarate@irit.fr

© The Editor(s) (if applicable) and The Author(s), under exclusive license   1
to Springer Nature Switzerland AG 2021
J. E. Hernández and J. Kacprzyk (eds.), *Agriculture Value Chain — Challenges and Trends
in Academia and Industry*, Studies in Systems, Decision and Control 280,
https://doi.org/10.1007/978-3-030-51047-3_1

1   Introduction

Information technologies offer a rich variety of tools to help practitioners make decisions. These tools, generally called Decision Support Systems (DSS) [1, 2], are available in multiple domains. In current times, these tools take the form of web-applications. Although some decisions a person makes may appear unconscious (such as those made while driving), other decisions require that the person stops, reflects, and selects what is considered the most effective alternative. The strategy a person uses depends on the importance of the decision. Intuitively, the more important the decision, the harder the decision maker needs to think about it. A complex decision requires that one applies specific knowledge, incorporates multiple perspectives, considers various sources of information, and carefully weights alternative courses of action. Given its complexity, this process often requires software support. Moreover, when decisions involve multiple stakeholders (as most decisions taken in the context of work) collaboration support becomes necessary. However, collaboration isn't always adequately supported.

In agriculture, decisions such as campaign planning, soil management, handling crop health, selecting crop varieties, and commercialization were once made on the basis of traditional practices. However, globalization and constant climate change have become a source of complexity, uncertainty, and risk that challenge traditional methods.

Farm Information Management Systems (FIMS) are software systems that support decision making in farms. As reported in a recent review [3], FIMS have evolved from the original simple record-keeping programs into sophisticated and complex systems to support production management. Current FMIS adopt and adapt features that are common to ERP systems, such as operations management, finance, inventory, reporting, machinery management, human resource management, sales, traceability, and quality assurance. In addition, they offer a set of specific features such as best practice, site management, and precision farming. FIMS were introduced in the 70s [4], and nowadays they are widely available. Nevertheless, FIMS adoption is still limited. In Brazil, for example, contract adjustments and farmers' experience were found to have a negative impact on adoption of FIMS [5]. A study that surveyed ICT adoption in New Zealand and Uruguay [6] proposed a model to understand adoption. The model states that farmer's attributes such as objectives, personality, education, learning style, and skills directly influence adoption. Therefore, the authors conclude that software developers must work with farmers; both during design, and later providing training and support. Moreover, they argue that systems need to configurable to suit a range of farmer characteristics.

Information is central to decision making [7]. Besides information available locally, which might be recorded with support of a FIMS, agriculture decision making requires timely information that is normally available elsewhere. The work presented in this article places emphasis in information available on the World Wide Web, as it is one of the most used sources of Big Data that is not always considered in the kind of systems described above. The Web is a constantly growing source of information, created by collaborating individuals around the world. However, most of the information published on the Web is isolated in silos, scattered across multiple sources, or unfit for use [8]. Moreover, as practice knowledge is naturally distributed among stakeholders, collaboration becomes paramount. During campaign planning, for example, farmers must evaluate different sort of seeds that can be used in the plot. They must consider the production history of the plot, the fruits or vegetables being produced in nearby plots, and the potential for the occurrence of plagues. This information gathering task requires to browse different websites where information about seeds is captured including vendor information, community reviews and success cases. To make matters more complicated, the activity if often performed collaboratively by farmers, agricultural engineers, and procurement agents without any collaboration tools.

In the context of the RUC-APS project [9], researchers have met with farmers in order to gather requirements for tools to support collaborative decision making based on information available on the web. In order to involve stakeholders from the start of the analysis and design phases, a design thinking approach was followed. Design thinking [10] is a set of strategies commonly used by designers, especially in the presence of wicked problems. Design thinking consists largely of three interrelated processed. Inspiration is the process that brings together designers and users. Ideation is the process in which designers and users collaborate to imagine, prototype and test solutions to key problems identified during the inspiration process. Solutions that pass the tests, are then implemented into products as part in the context of the Implementation process.

This article presents a case study that shows how design thinking, collaboration engineering patterns, and Model-Driven Web Augmentation may be combined to produce a Minimal Viable Product (MVP) for collaborative campaign planning with information from the web. The rest of this document is organized as follows: Sect. 2 provides background. Then, Sect. 3 presents an overview of the approach. Section 4 shows a comprehensive example. And Sect. 5 concludes and talks about the future works.

2 Background

This section provides context to the information discussed throughout the research paper. The Background includes a brief description of main building blocks of the approach followed.

2.1 Collaboration Engineering and Patterns

Systems designers have learned the value of documenting, in the form of design patterns, proven solutions to recurring problems [11]. Design patterns increase productivity, bring quality, and support effective communication among designers and developers. Patterns exist to support and speedup the design of computer mediated collaboration [12]. Researchers observed that, in effective collaborative processes, participants engage in recurrent patterns of collaboration [13]. There is a "diverge" pattern of collaboration characterized an increase in the number of available choices (e.g., ideas). It follows a "converge" pattern of collaboration where participant engage to obtain less choices of higher value. These choices are then "organized", as participants get a better idea about the relationships among choices. The resulting choices are "evaluated" to better understand how each of them contributes to the goal. To complete the cycle, participants engage to "build consensus".

Researchers also observed that the success of these collaboration initiatives was strongly related to the availability of a facilitator whose role is to effectively guide participants though the process. On the basis of this understanding, they proposed Collaboration Engineering [13–15]. Collaboration engineering provides a series of strategies and tools that capture effective collaboration strategies, in a way that makes results replicable. A key element in collaboration engineering is the ThinkLet. A ThinkLet is packaged facilitation intervention that creates a predictable, repeatable pattern of collaboration among people working together toward a goal [15]. ThinkLets consist of three components: (a) a tool (e.g. a software); (b) a configuration for the tool; (c) a facilitation script. The creators of this methodology published a catalog including detailed description of 40 ThinkLets. However, the methodology allows for new ThinkLets to be discovered and recorded.

2.2 Design Thinking

Design thinking [10] is a set of strategies commonly used by designers, especially in the presence of wicked problems. It is widely used to foster innovation. Although there are various models to describe the design thinking process, it largely consists of three interrelated processes or activities: Inspiration, Ideation, and Implementation. As a general rule, these activities are seen as a phase of divergent thought followed by a phase of convergent thought.

Inspiration is the activity that brings together designers and users. The goal of inspiration is to identify, understand, and record key challenges or problems. Empathy and collaborative, multidisciplinary work are key elements during this activity. Multiple tools and strategies can be used to support this activity, for example Journey Mapping.

Ideation is the process in which designers and users collaborate to imagine, prototype and test solutions to key problems identified during the inspiration process. Many proposals will be produced, only a few of which will have a change of being discussed with the end customer. Designers must turn to prototyping techniques that require low effort (for example paper mockups). However, there is a trade-off between low effort, and the ability of the "low-fidelity" prototypes to convey the solution idea with enough clarity. For the case of software solutions, the ability to interact with the prototype can make a big difference. Model-driven web augmentation, and collaboration patterns are tools that can support the ideation activity.

The prototypes that pass the ideation phase have a chance of becoming final products. During the implementation activity, stakeholders decide how much effort they will invest to develop the selected ideas. As they still don't know how well they will perform, they may decide to start small and test, for example, in pilot scale. During implementation, model driven web-augmentation can also play an important role.

2.3   Group Decision Support Systems

Increasingly challenging global and environmental requirements have resulted in agricultural systems coming under increasing pressure to enhance their resilience capabilities. As stated by Hernandez et al. colleagues [16], integrated solutions are necessary to support knowledge-management, collaboration, risk management and regulation management across agriculture stakeholders. Such solutions need to consider both the capabilities of the web as a source of reusable information, and the web as platform on which systems (applications) operate. As agricultural production systems are rich networks of independent actors, collaboration becomes paramount.

Group Decision Support Systems are software systems designed to help multiple stakeholders during decision making processes. They offer support for communication, collaboration, coordination, information retrieval and, in some cases, incorporate advances in operational research regarding the solution of multi-variable problems. Different GDSS emphasize different perspectives of the challenge. Some of them accentuate the importance of problem modeling, some of them focus on facilitating collaboration among stakeholders and some others adopt a MultiCriteria approach [17].

2.4   Web Augmentation

Since almost twenty years, hypermedia and web communities are dealing with the alteration and integration of existing third-party applications, which was coined as Web augmentation [18, 19], and currently this alteration is mostly performed be means of Web browser extensions. When Web browsers reached widespread use and social networks started to arise, also the first actual communities around software for adapting existing Web sites emerged. For instance, an important community emerged around *userscripts*, which are JavaScript script that are installed and executed on the browser when a Web page loads, allowing its alteration.

Nowadays, the techniques behind Web augmentation are used by a vast number of users that may choose among thousands of extensions for adapting Web content may be found at the Web browser stores, and significant developer communities support some of these tools. For instance, the Userstyles community (https://userst yles.org) offers a wide number of scripts that augment Web sites by adding further CSS specifications that change the content presentation. Userscripts communities, such as Greasyfork (https://greasyfork.org/), offer repositories of scripts with a wider spectrum of purposes, since they support different weavers of JavaScript code (e.g. GreaseMonkey or TamperMonkey); therefore, it is possible to change not just the style but the content and behavior of a Web page.

End-users and other stakeholders with programming skills may interact in such communities for the creation, sharing, and improvement of specific augmentation artifacts [20]. In all these communities, no matter which tool they support, there is a dependency between users with and without programming skills, since not all of them can implement the solutions they need and ask others for help. In this light, some research works proposed End-User Development (EUD) approaches to let users specify their own augmentation artifacts; these works are discussed in the related work section [21, 22].

Although the existence of EUP environments makes the alteration of UI straightforward, complex requirements such as collaboration, require more than just standalone client-side scripts [18]. In this sense, involving a method to integrate back-end counterparts (i.e., a web application working on a server) is needed. Still with the focus on fast prototyping, it is well-known that MDD approaches are more productive, therefore, this work proposes a Model Driven Web Augmentation approach, which is based on the models at both sides, front-end and back-end.

3 The Approach in a Nutshell

Existing software engineering practices establish the identification of system requirements as the starting point. Analysts record what the prospective users expect the system to do on their behalf. This practice assumes that the prospective users have a concrete list of requirements that need to be fulfilled, and that they understand what a software system can do for them. Agriculture, however, is a domain in which penetration of modern Web technologies is low. Apart from early adopters and experimental programs, most farmers are not aware of the potential that information technologies can offer. In this context, traditional requirements elicitation techniques do not apply, and systems design becomes an exploratory and iterative innovation process that calls for adequate tools. Nevertheless, if a set of Web applications related to

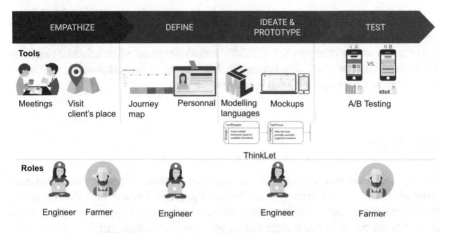

Fig. 1 The design thinking process based on MDWA

the desired domain is available, they could be used as a basis to speed up information requirements gathering. In this way, by augmenting these Web applications new requirements may be considered and modeled.

Previous work [23] introduced the foundations of an approach for modeling complex augmentation that comprises both client and server-side artefacts. The approach takes advantages of model-driven claimed benefits, such as high-productivity and less error-prone than traditional development approaches that start from scratch.

This work, presents an approach giving support to complex business process comprising inter-site information and groupware requirement gathered from end-users. It relies on a design thinking approach for extracting and prioritizing requirements and uses Model-Driven Web Augmentation technologies for producing an MVP (Minimum Viable Product).

This work is based on the design thinking approach proposed by Hasso-Plattner Institute of Design at Stanford which considers 5 steps: Empathise, Define, Ideate, Prototype, and Test. Figure 1, introduces the main steps of the approach and related artifacts. In the image tools used in the approach as well as players who perform the activities are shown.

3.1 Empathize

The Inspiration step focuses on understanding the end-user situation and identifying the opportunities to improve the business experience. The information gathering requires visiting end-user's place, interview stakeholders, and perform a business activity on behalf of the end-user. In this way, analysts can experience and explore the whole business context.

In order to get the most from the interviews, the analyst does not only focus on the business processes and economic goals of the business, indeed he must maximize his empathy understand psychological and emotional needs of stakeholders and thus understand what is done, why is done in that way and how stakeholders feel about the experience highlighting what is really meaningful.

3.2 Define

The problem statement is defined in this step. The team describes the issues that need to be addressed in each activity of the business process. The captured understanding during meetings and visits can be documented using the journey map tool which describes the business experience points, pains, opportunities, etc.

In this stage the co-creative tool Journey Mapping is used for gathering processes information which is developed jointly with stakeholders. One of the requirements is the definition of the personas profile used for depicting the main business actors. According to the experience exposed in this step, the process activities are ranked somehow it is clear if it has a positive or negative contribution to the overall experience. Additionally, the pains are outlined and opportunities are presented. The resulting Journey Mapping document along with personas' profiles are later during the ideation step.

3.3 Ideate and Prototype

The ideation is a twofold process where firstly the collaboration requirements are analyzed and designed, and secondly these are modeled using a Model-Driven Web Augmentation Approach.

Collaboration Engineering
As people move together towards a common goal (e.g., to make an informed decision) they engage in a process in which they contribute with their knowledge, effort, and resources. This collaborative process can be characterized as consisting of the following phases: understanding the problem, developing alternative solutions, evaluating alternatives, choosing alternatives, making a plan, taking action, and monitoring results.

Once the journey map is elaborated, a throughout analysis of pains is performed in such way collaboration opportunities are identified. By using ThinkLets tool, the software engineers ideate improvements in the existing business process.

A Model-Driven Web Augmentation Approach

A Model-Driven Web Augmentation approach is used by Software engineers to develop complex inter-site by designing both client and server-side augmentation components. The resulting integration seeks to accomplish business goals through the implementation of business processes. The approach for designing and developing Web augmentations requires following steps:

- Decouple the augmentation from the core application by introducing a design layer (called Augmentation Layer), which comprises additional conceptual, navigational an interface models apart from the ones of the core application. This layer groups all the new elements that will enhance existing Web sites that are used in the business process being augmented.
- Capture the basic conceptual model by tagging the information presented in the core application pages, in a similar fashion to WOA [24]. The lack of access to the underlying business models, which models business entities and their attributes, of the applications being augmented requires the generation of a second conceptual model; the one inferred by the user perception of the augmented. In this process, data elements in the page are tagged and grouped by an augmentation analyst into an entity definition. In such a way, a simplified conceptual model is obtained. The augmentation analyst is a skilled end-user with advanced knowledge of Web Applications, whose goal is to improve a core application. In further steps, the model instantiation in a particular user session will be used for giving contextual information to the augmentation engine by providing model instances information when triggering the augmentation.
- Augmentation requirements are modeled using Web engineering notations (e.g. use cases or user interaction diagrams) and separately mapped onto the following models using the heuristics defined by the design approach (e.g. [25]). Notice that, the augmentation requirements are not integrated into the core requirements model, leaving their integration to further design activities.
- New behaviors, i.e. those belonging to the Augmentation layer, are modeled as first-class objects in the augmentation conceptual model. Such model defines all the objects and behaviors corresponding to the new requirements. Additionally, it may include the core application conceptual classes perceived by the augmentation analyst, allowing defining relationships between the augmentation business model and the core application. Notice that this strategy can be applied to any object-oriented method, i.e., any method using a UML-like specification approach.
- Nodes and links belonging to the augmentation's navigational model may or may not have links to the core navigational model. The core navigational model is also oblivious to the augmentation's navigational classes, i.e., there are no links or other references from the core to the augmentation layer. This principle can be applied in any Web design approach.

- This approach uses a separate specification for the connection between the core and the augmentation nodes. As will be shown later in the paper, the integration is achieved at run-time as part of a client-side weaving engine. Conversely, in other model-driven approaches, the integration can be performed during model transformation which is not the case of this approach.
- Optionally, the interfaces corresponding to both the application being augmented and the new content expected to be introduced are designed (and implemented). For such task engineers rely on Mockups which are used to document how User Interfaces should look like by example.

In [23] was presented the usage of Web augmentation to give support to the evaluation of raw material used during the vegetables production.

Testing
Design thinking proposes a flexible testing session that allows to change the prototype on-the-fly during user tests if this is required. The MDWA approach presented here provides a very convenient environment for this kind of activity. First, the client-side component allows to redefine the augmentation weaving on-the-flight, which is made visually on the browser. Second, using EUD tools for Web augmentation, the UI may be easily. Third, if the augmentation process requires some adaptation on the back-end component, since it is deployed using a MDD approach and cloud services, it may be also altered during the user test session.

4   Running Example

The green belt of La Plata city is a farming area of approximately 6.000 ha that provides fresh vegetables to a large part of the population of the Buenos Aires province, Argentina. Tomato and green leafy vegetables are key product of the area. According to the last available census data [26] there were in 2005 over 1000 farms in this area. This section describes how the approach was applied in a mid-size farm in La Plata city. The experience was sponsored by farm's managers and owners giving access to the facilities and promoting meetings with farm's employees.

4.1   Empathize

Multiple meetings with farmers, agronomists, and experts were conducted in the green belt of the La Plata city. In all these meetings a recurring problem was the difficulty to assess the plans of the farming community as a whole and consequently adapt one's own farming plan. In Figs. 2 and 3, show pictures of meetings celebrated at farm's facilities and at a research lab; respectively.

As a result, a farmer that decided to plant 10 ha of tomato, could learn near the time of the harvest that there was an over production of tomato and the price has

Fig. 2 The farm's manager describes some challenges of producing tomatoes

Fig. 3 Software engineers interview to farm's stakeholders

substantially dropped. There is no up-to-date information regarding the production in this area on which farmers can base their decisions. There is no central organization that helps farmers plan production. Consequently, when farmers make a decision that depends on the projected production of a given crop, they do this by resorting to intuition and talks with a few colleagues.

4.2 Define

In Fig. 4 the resulting journey map is shown (see[1] for further detail). Most important production steps are analyzed by Software engineers who apply collaboration lens to identify opportunities. As journey maps list a myriad of relevant challenges and improvement opportunities, the software engineer agree with farmers on prioritizing

---

[1]Analysis of fresh vegetables producer—a journey map https://bit.ly/2Ku6Jyl.

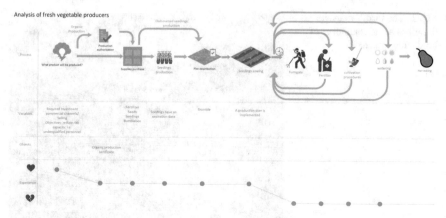

Fig. 4 Journey map of vegetables producers

and choosing the need to work on because it can be found opportunities from robots to assist manual works relating to harvesting to cultural works training for employees.

4.3 Ideate and Prototype

This step shows how collaboration engineering and MDWA can help produce low effort prototypes that feed-back into the innovation cycle.

Engineering the Collaboration Between Farmers

Regardless of the higher quality of the tomatoes produced by the farm members, they were not able to obtain a good market price and make a profit. This year they want to plan better and try to predict possible scenarios or outcomes and plan accordingly. They know "everything is on the internet now" so, with the help of a group collaboration engineers they prototype the tools they need to harness the power of the web.

Together, they design a collaboration process facilitated by ThinkLets implemented on top of web augmentations. The diagram shown in Fig. 5 using ThinkLet Modelling Language provides an overview of the collaboration process they designed.

Next the collaboration scenario is characterized:

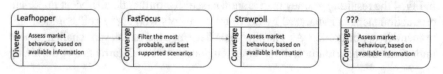

Fig. 5 ThinkLet chain of collaboration features

Who Members of a community of familiar farmers. They share knowledge to improve everyone's results. They share and exchange their production surplus.

Objective Collaborate during the planning of the upcoming season in order to help each other find the best seeds/seedlings, and other required inputs, and possibly obtain a better price by buying as a group.

Context The group has a Facebook page. They normally use Facebook (posting in the group, messaging) to share information. However, information cannot be organized, ranked, related and turned into plans. "An app to support communities like us would be a total success" they said. They looked for an "app" they can use to plan the season, but nothing seems to be adequate. Therefore, they met with a team of IT guys to create it. Luckily, they chose a design thinking approach, collaboration engineering patterns, and web-augmentation to build a first prototype. This is what they came up with.

The next section depicts how a model driven approach can speed up the prototype development.

Augmentation Development

Previous work [27] introduced the definition of Concern-Sensitive Navigation (CSN) for a Web application when the content, links, and operations exhibited by the core application pages are not fixed for different navigation paths, but instead can change when accessed in the context of various navigation concerns. In CSN, its properties can be slightly adjusted according to the current user's concern; for example, for assisting the vegetable production planning. Web augmentation can be used to implement CSN in Web applications because of the applications' pages are enhanced with new content, navigation, and links according to the task purpose; supporting the underlying user's activity that motivates the navigation. As a result, the user's navigations may help to improve the user's experience while performing the activity.

Considering the scenario described early in this section in the domain of agriculture: planning a vegetable production by a farmer. One of the minor tasks of the planning activity is the research of different products alternatives used during the vegetable production.

As part of the ideation step stakeholders proposed the Far-o-matic tool which is a digital facilitation tool that guides farmers along a multi-step process. The first step was to identify the community needs and the best alternatives to cover them. In the second step, farmers build their plans choosing inputs from the best alternatives already identify. At this point they indicate how much they need and what they will produce. The result is an overview of the potential yields for the whole community and total required inputs. This report is the focus of a shared group discussion, that might drive some farmers to adjust their plans. They might step two and three a couple of times. At the final step, each of them prints a report. The reports include not on-line the plans, but also the shopping cart that can both be used for a common purchase or for individual purchases. The main features of the tool being modelled are:

Start The tool is built on top of Facebook. Whenever farmers log in to Facebook, the tool adds an additional menu entry: Farm-o-matic.

The Wish List First, each of them (with a plan in mind) collects products from various web-sites. Products are saved to a shared wish-list. Collected products are visible for everyone. It is possible comment in favor, or against a product. In addition, one can offer an alternative for a product (it can be the same product from a different provider and/or different price, or an alternative product). The result is a list of products, and potential alternatives for some of them.

Sending the Best Alternatives to the Shopping Cart Some products in the list have alternatives. Farmers vote, in each case, which alternative they recommend. They can vote only one per group of alternative products. When voting finished, all products are transferred automatically to the shopping cart. For those products that had alternatives, only the most voted one goes to the cart.

The Purchase Order Now, farmers express their purchase plans, indicating how much of each product in the shopping cart they need. At all times they can browse and print the global purchase order, and the individual orders.

In order to support abovementioned requirements some design tasks need to be carried out.

Extracting Conceptual Model from Core Application
In the approach presented here, augmentation functionalities might be new behaviors which are added to the conceptual model (and which might encompass many classes). Figure 6 presents the conceptual model composed by three entities:

- Seed to capture alternative species of a given product.
- Pesticide to identify the requirements and constraints of the products and the vegetables.
- Fertilizer to identify the product and used technique.

This model is used for persisting a pre-selection of Seeds, Pesticides and Fertilizers that will be evaluated to define which vegetable will be produced by a farmer.

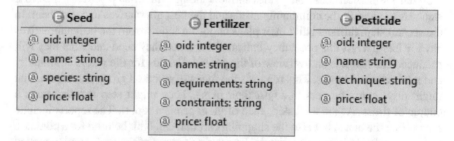

Fig. 6 Conceptual model of products being reviewed by farmers

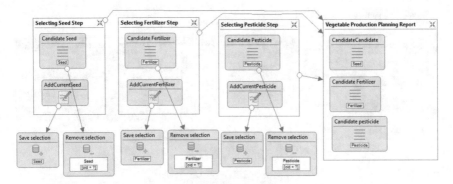

Fig. 7 Navigation model depicting how the product selection activity behaves

Modeling the Augmentation Layer

This example introduces an interesting challenge of aggregating information gathered from the navigation of several sites and its persistence. Therefore, the backend component plays a primary role since it allows to centralize the user's activity information storing the navigation's state defined by navigating several Website.

Figure 7 presents the navigation model required for supporting the products' information gathering task. The model is based on the IFML standard for describing how the application will be navigated. Each time a user accesses a Web site that contains a Seed, a Pesticide, or a Fertilizer concept, a bucket (Selecting Seed, Selecting Fertilizer, and Selecting Pesticide) of such entity lists the product's details (the concept attributes, such as the name or the price). Besides, it is possible to add the current visited product to the list or delete one product from the list. At any time, the end-user can review current task advance by accessing planning report page through a link present in each bucket. The flow from Selecting Seed, Selecting Fertilizer, and Selecting Pesticide pages to the Report one allows such navigation.

The result of enriching a Web site with features that allow the user to gather information, which will be used later, during a complex process that involves several activities in different Web sites is shown in Fig. 8. The first step is the Seed selection at Lowe's.[2] Then, several fertilizers are selected by a farmer user to be evaluated picking them from Amazon.[3] As a third step, a support for saving a list of pesticide is added to the TractorSupply[4] web site honoring. The user can access to a summary page where all gathered information is shown and can be used to make a decision for the vegetable to be produced. Note that the navigation must not be performed necessarily in the presented order. Indeed, the user can browse all the applications (back and forth) until he obtains the information he needs for making a decision.

The WebRatio [1] tool supports the IFML approach and the designers can use the tool for modeling the business model and the navigational. The tool generates the Java

---

[2]https://www.lowes.com, last accessed July 27th 2017.

[3]https://www.amazon.com, last accessed July 27th 2017.

[4]https://www.tractorsupply.com, last accessed July 27th 2017.

Production planning

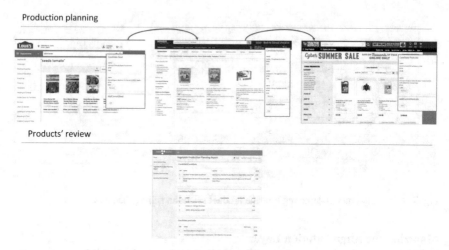

Products' review

Fig. 8 Final augmentation results and gathered information report

code related to the modelled entities and relationships as well as the modelled Web application. Moreover, the tool generates the ORM descriptors for Hibernate framework which allows persisting and retrieving objects from/to most of the database engines (i.e. MySQL, Microsoft SQL and Oracle). This approach promotes the use of MDWE approaches for modelling Web Augmentation because the benefits reported on developer's performance. However, the Web augmentation client and server-side artifacts can be developed from the scratch based on software engineering practices.

5 Conclusions

Decision making in agriculture is as complex as it is in other domain because of the variables involved in any decision and their relationship. The information and tools available in the Web are key factors nowadays to make better decisions as they are used by main agriculture actors in for example planning activities. This article presented a novel approach which relies on different techniques for assisting the user's needs (traditional known as requirement gathering task) based on exploratory studies and empathy task as well as generating solutions built on top of existing Web apps. By combining Web Augmentation techniques with Model-Driven Web Engineering approaches, the solution only requires to be idealized and designed and source code is generated without requiring a software developer intervention. To illustrate the approach, introduced a case study which involved performing a design thinking experience in a farm located in La Plata, Argentina was introduced. The outcome of the study was a tool that helps farmers, and procurement agents to gather information of raw material, collaboratively evaluate the alternatives and finally assist the decision making task.

Innovation in information systems for agriculture is needed. Information systems have the potential to help stakeholders deal with the stressing conditions imposed by climate and market change, and to embrace sustainable practices. However, innovation is a domain where information systems are still maturing (such as agriculture) is challenging. This work showed that Design Thinking, Web Augmentation, and Collaboration Engineering are effective tools to foster innovation in agriculture.

Further work involves conducting an experiment to assess the effectiveness and productivity of this approach. There are also plans to extend the scope of the application to other business domain such as financial and retail.

Acknowledgement Authors of this publication acknowledge the contribution of the Project 691249, RUC-APS: Enhancing and implementing Knowledge based ICT solutions within high Risk and Uncertain Conditions for Agriculture Production Systems (www.ruc-aps.eu), funded by the European Union under their funding scheme H2020-MSCA-RISE-2015.

References

1. Sprague, R.H., Jr., Carlson, E.D.: Building Effective Decision Support Systems. Prentice Hall Professional Technical Reference, Upper Saddle River (1982)
2. Marakas, G.M.: Decision Support Systems in the 21st Century, vol. 134, 2nd edn. Prentice Hall, Upper Saddle River (2003)
3. Fountas, S., et al.: Farm management information systems: current situation and future perspectives. Comput. Electron. Agric. **115**, 40–50 (2015)
4. Blackie, M.J.: Management information systems for the individual farm firm. Agric. Syst. **1**(1), 23–36 (1976)
5. Carrer, M.J., de Souza Filho, H.M., Batalha, M.O.: Factors influencing the adoption of farm management information systems (FMIS) by Brazilian citrus farmers. Comput. Electron. Agric. **138**(C), 11–19 (2017)
6. Alvarez, J., Nuthall, P.: Adoption of computer based information systems: the case of dairy farmers in Canterbury, NZ, and Florida, Uruguay. Comput. Electron. Agric. (2006)
7. Simon, H.A.: The new science of management decision. Prentice Hall PTR, Upper Saddle River (1977)
8. Wang, R., Strong, D.: Beyond accuracy: what data quality means to data consumers. J. Manage. (1996)
9. Hernandez, J.H.E., et al.: RUC-APS: enhancing and implementing knowledge based ICT solutions within high risk and uncertain conditions for agriculture production systems. In: 11th International Conference on Industrial Engineering and Industrial Management (2017)
10. Brown, T.: Change by Design (2009)
11. Gamma, E., Helm, R., Johnson, R., Vlissides, J.: Design Patterns: Elements of Reusable Object-Oriented Software. Addison-Wesley Longman Publishing Co., Inc., Boston (1995)
12. Schummer, T., Lukosch, S.: Patterns for Computer-Mediated Interaction. Wiley, Hoboken (2013)
13. Briggs, R.O., De Vreede, G.-J., Nunamaker, J.F., Tobey, D.: ThinkLets: achieving predictable, repeatable patterns of group interaction with group support systems (GSS). In: Proceedings of the 34th Annual Hawaii International Conference on System Sciences, 9 pp (2001)
14. Lowry, P.B., Nunamaker, J.F.: Using the thinkLet framework to improve distributed collaborative writing. In: Proceedings of the 35th Annual Hawaii International Conference on System Sciences, pp. 272–281. HICSS (2002)

15. Briggs, R.O., De Vreede, G.-J., Nunamaker, J.F., Jr.: Collaboration engineering with ThinkLets to pursue sustained success with group support systems. J. Manage. Inf. Syst. **19**(4), 31–64 (2003)
16. Hernandez, J.E., et al.: Challenges and solutions for enhancing agriculture value chain decision-making. A short review BT. In: Collaboration in a Data-Rich World, pp. 761–774 (2017)
17. Zaraté, P., Camilleri, G., Kilgour, D.M.: Multi-criteria group decision making with private and shared criteria: an experiment. In: International Conference on Group Decision and Negotiation, pp. 31–42 (2016)
18. Bouvin, N.O.: Unifying strategies for web augmentation. In: Proceedings of the Tenth ACM Conference on Hypertext and Hypermedia: Returning to Our Diverse Roots, pp. 91–100 (1999)
19. Díaz, O., Arellano, C.: The augmented web: rationales, opportunities, and challenges on browser-side transcoding. TWEB **9**(2), 8 (2015)
20. Firmenich, D., Firmenich, S., Rivero, J.M., Antonelli, L., Rossi, G.: CrowdMock: an approach for defining and evolving web augmentation requirements. Requirements Eng. **23**(1), 33–61 (2018)
21. Díaz, O., Arellano, C., Aldalur, I., Medina, H., Firmenich, S.: End-user browser-side modification of web pages. In: Web Information Systems Engineering—WISE 2014, pp. 293–307 (2014)
22. Aldalur, I., Winckler, M., Díaz, O., Palanque, P.: Web augmentation as a promising technology for end user development. In: Paternò, F., Wulf, V. (eds.) New Perspectives in End-User Development, pp. 433–459. Springer International Publishing, Cham (2017)
23. Urbieta, M., Firmenich, S., Maglione, P., Rossi, G., Olivero, M.A.: A model-driven approach for empowering advance web augmentation—from client-side to server-side support. In: APMDWE (2017)
24. Firmenich, S., Bosetti, G.A., Rossi, G., Winckler, M., Barbieri, T.: Abstracting and structuring web contents for supporting personal web experiences. In: Web Engineering—16th International Conference, (ICWE) 2016, Lugano, Switzerland, 6–9 June 2016. Proceedings, pp. 77–95 (2016)
25. Popovici, A., Gross, T., Alonso, G.: Dynamic weaving for aspect-oriented programming. In: Proceedings of the 1st International Conference on Aspect-Oriented Software Development, pp. 141–147 (2002)
26. Censo hortifloricola de la provincia de buenos aires (2005)
27. Firmenich, S., et al.: Engineering concern-sensitive navigation structures, concepts, tools and examples. J. Web Eng. **9**(2), 157–185 (2010)

A Conceptual Framework for Crop-Based Agri-food Supply Chain Characterization Under Uncertainty

M. M. E. Alemany, Ana Esteso, A. Ortiz, J. E. Hernández, A. Fernández, A. Garrido, J. Martín, S. Liu, G. Zhao, C. Guyon, and R. Iannacone

Abstract Crop-based Agri-food Supply Chains (AFSCs) are complex systems that face multiple sources of uncertainty that can cause a significant imbalance between supply and demand in terms of product varieties, quantities, qualities, customer

M. M. E. Alemany (✉) · A. Esteso · A. Ortiz
Universitat Politècnica de València, Camino de Vera s/n, 46022 Valencia, Spain
e-mail: mareva@omp.upv.es

A. Esteso
e-mail: aneslva@doctor.upv.es

A. Ortiz
e-mail: aortiz@upv.es

J. E. Hernández
University of Liverpool Management School, Liverpool, UK
e-mail: J.E.Hernandez@liverpool.ac.uk

A. Fernández · A. Garrido · J. Martín
LIFIA, Facultad de Informática, Universidad Nacional de La Plata, 50 y 115 s/n,
1900 La Plata, Argentina
e-mail: alejandro.fernandez@lifia.info.unlp

A. Garrido
e-mail: garrido@lifia.info.unlp.edu.ar

J. Martín
e-mail: jonathan.martin@lifia.info.unlp.edu

A. Garrido
CONICET, Buenos Aires, Argentina

S. Liu · G. Zhao
Plymouth Business School, University of Plymouth, Drake Circus, Plymouth PL4 8AA, UK
e-mail: shaofeng.liu@plymouth.ac.uk

G. Zhao
e-mail: guoqing.zhao@plymouth.ac.uk

C. Guyon
Bretagne Development Innovation, 1bis Route de Fougères, 35510 Cesson-Sévigné, France
e-mail: c.guyon@bdi.fr

19

J. E. Hernández and J. Kacprzyk (eds.), *Agriculture Value Chain — Challenges and Trends in Academia and Industry*, Studies in Systems, Decision and Control 280,
https://doi.org/10.1007/978-3-030-51047-3_2

requirements, times and prices, all of which greatly complicate their management. Poor management of these sources of uncertainty in these AFSCs can have negative impact on quality, safety, and sustainability by reducing the logistic efficiency and increasing the waste. Therefore, it becomes crucial to develop models in order to deal with the key sources of uncertainty. For this purpose, it is necessary to precisely understand and define the problem under study. Even, the characterisation process of this domains is also a difficult and time-consuming task, especially when the right directions and standards are not in place. In this chapter, a Conceptual Framework is proposed that systematically collects those aspects that are relevant for an adequate crop-based AFSC management under uncertainty.

Keywords Crop-based agri-food supply chain · Conceptual framework · Uncertainty · Management

1 Introduction

The term "agri-food supply chain" (AFSC) has been defined as a set of activities necessary to bring agricultural products "from farm to fork" [1–8]. Thus, AFSCs are responsible for the production and distribution of both vegetable and animal-based products [9]. The production processes of meat and horticulture sectors are very different. For this reason, different frameworks are needed for characterizing them. Since horticulture sector (crop-based AFSCs) has received the least attention in the literature, this chapter will center the research on it.

Crop-based AFSCs are complex systems that face multiple sources of uncertainty that can cause a significant imbalance between supply and demand. Poor management of these sources of uncertainty can have negative impact on quality, safety, sustainability and logistic efficiency of the products and processes as well as in the waste. Thus, a mandatory previous step to develop any decision support tool, especially for the management of every supply chain, is to define the problem under study very precisely, i.e. in a structured way and in a natural language understandable for every crop-based AFSCs implied stakeholder.

The characterisation process of this domains becomes a difficult and very time-consuming task, especially when standards are not in place. Along these lines, the main objective of this paper is to identify the relevant and distinguishing characteristics of crop-based AFSCs and their sources of uncertainties as a first step in the proper management of different SC processes with the support of different technologies. This is a necessity that arises in the context of the European Project RUC-APS 691249 [10] where, before developing any solution, it was necessary to achieve an understanding between academics and non-academics in order to define agilely

R. Iannacone
Agenzia Lucana di Sviluppo e di Innovazione in Agricoltura, Centro Ricerche Metapontum Agrobios, Metaponto di Bernalda (MT), Bernalda, Italy
e-mail: rina.iannacone@alsia.it

the problem to be solved. Conceptual Frameworks have proven their utility and usability for addressing this challenge. Miles and Huberman [11] define a Conceptual Framework as a visual or written product that explains in a graphical or literary way the elements to be studied, the key factors, the concepts or variables and their relationships.

In this paper, a novel Conceptual Framework is proposed that systematically collects relevant aspects for the proper management of crop-based AFSC in an uncertain context based on five views of the system (Physical, Functional, Organization, Informational and Decisional) that, to the best of our knowledge, have not been previously considered in an integrated way. This Conceptual Framework offers several advantages. Firstly, the Conceptual Framework constitutes a tool for the understanding among academics and non-academics in order to precisely define the problem under study. Secondly, it could be used as a reference model for the subsequent development of particular models to support crop-based AFSCs decision-making under uncertainty. Thirdly, the Conceptual Framework could be used to review existing approaches in the literature in a structured way. This can help practitioners when searching already existing solutions and can support researchers for identifying gaps in the topic.

The rest of the paper is structured as follows. First, the main blocks to be integrated in the Conceptual Framework are presented. Then, different sections are dedicated to each block, containing detailed description of the constituent elements. Finally, some conclusions are outlined.

2   Conceptual Framework

The Conceptual Framework proposed in this paper is structured in different views (Fig. 1), similarly to that proposed by Alemany et al. [12] but with differences regarding the elements of each view and its content.

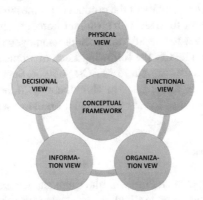

Fig. 1 Views integrated in the conceptual framework for crop-based AFSC

We assume that crop-based AFSCs decisions (Decisional View) are made on elements such as items and physical and human resources (Physical View), which are specifically arranged (Organisation View), and that particular information (Information View) is required to properly make decisions. The elements of each View and their relationships are described below.

3 Physical View

The definition of the problem requires to delimit the physical scope under study, that is, the part of the AFSC that is under our influence as well as their characteristics. The Physical View identifies how a specific supply chain is configured (designed) taking into account its nodes, resources and products including their flows through the supply chain.

3.1 Products

One of the main differentiating aspects of crop-based AFSC is the intrinsic characteristics of their products. Based on them, different classification of products can be found in crop-based AFSC literature. Van der Vorst et al. [9] and Teimoury et al. [13] differentiate between products which are going to be directly consumed and products which are going to be processed before being consumed. The formers are also classified, in turn, in perishable and non-perishable products [14]. Differently, Zhang and Wilhelm [15] state that agricultural products may be classified as field crops (e.g., corn, cotton, rice, seeds and wheat) or specialty crops (fruits, vegetables, grapes and wine, ornamentals, tree nuts, berries and dried fruits).

On the other hand, Grillo et al. [16] identify the AFSCs as one of the Lack of Homogeneity in the Product sectors. Lack of Homogeneity in the Product is characterized by the heterogeneity of the products in some attributes that are relevant to the customers. In order to meet with customer homogeneity requirements, these SCs include sorting stages to classify products into homogeneous subtypes based on certain attributes. They identify the following Lack of Homogeneity in the Product characteristics also valid for crop-based AFSCs: Subtype, Subtype Quantity, Subtype Value and Subtype State. These Lack of Homogeneity in the Product features are used to characterize the crops for our Conceptual Framework proposal (Fig. 2).

3.1.1 Subtype Number

Subtypes are units of the same crop with the same characteristics requested by the customer. The classification of horticulture products into subtypes depends on the defined attributes to sort products and their possible values. For instance, Blanco et al.

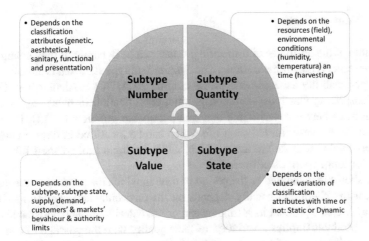

• Depends on the classification attributes (genetic, aesthtetical, sanitary, functional and presenttation)

• Depends on the resources (field), environmental conditions (humidity, temperatura) an time (harvesting)

Subtype Number

Subtype Quantity

Subtype Value

Subtype State

• Depends on the subtype, subtype state, supply, demand, customers' & markets' bevahiour & authority limits

• Depends on the values' variation of classification attributes with time or not: Static or Dynamic

Fig. 2 Products' characteristics of crop-based AFSC

[17] state that during the production process there are different classification stages that provide with a large quantity of final products based on the specific combination of variety, quality, weight, size and packaging. Grillo et al. [18] define subtypes of tarongines based on variety, quality, calibre, packaging type, and harvesting time. Verdouw et al. [19] identify the main criteria for the selection and classification of fruit: the size, weight, maturity, damage, deterioration, color, shape and ripeness.

Therefore, the sorting attributes of crops can be considered genetic (varieties), sanitary (fungi), aesthetic (damages, color, caliber, etc.), functional (sugar content related to taste, freshness) and based on presentation (packaging type). The specific sorting attributes and their potential values will depend on the specific crop provided. Finally, the possible combination of the values of a variety of attributes will provide the AFSC with the number of existing subtypes, which is usually variable and uncertain.

3.1.2 Subtype Quantity

Although the final quantity obtained of each subtype may depend on the lot size, its proportion is usually variable. Verdouw et al. [19] states that there is a variation of the quality and, therefore, an appearance of subtypes, between different producers, different production lots and even in the same lot. These factors can depend on the resource (field), the moment of time (harvesting time) and the environmental conditions (temperature and humidity) that increase the diversity and the appearance of subtypes due to the perishable nature of the products. The input–output relationship between the amount of input to a classification stage and the classified output, can be estimated based on historical percentages, but it can be said that it is variable and uncertain.

3.1.3 Subtype State

The value of the attributes of a specific product can be dynamic, if they change over time, or static, if they remain stable. The dynamic subtype state that is closely related to the perishability aspect is one of the crop-based AFSCs particularities. Thus, it is not surprising that most common product classification in these supply chains distinguishes between perishable and non-perishable products [3, 13]. Perishable products have a very limited shelf life which can be measured in days. In contrast, non-perishable products are those which, while having a limited shelf life, can be stored for longer periods of times [3].

The shelf life is defined as the period of time in which the product ceases to have value for the client and usually depends on the environmental conditions. During the shelf-life of a product its attributes can or not deteriorate. The perishable aspect of fresh products implies a variation of their quality with the passage of time [20]. Therefore, it can be said that the state of the subtype is dynamic. Nahmias [21] establishes a classification of perishable food into two categories depending on shelf life:

- Fixed shelf life: the shelf life of the item is predetermined and once it has elapsed, the item becomes worthless: e.g., salads prepared ready-to-eat. When the shelf life is fixed, the most common visual clue is "preferentially consume before" (best-before-date). In this case, customers adapt the price they are willing to pay based on how far the best-before-date is.
- Random life: there is no pre-established length for the shelf life of these items. Their length is random and the probability distribution can take several forms (e.g. vegetables). In this case, the expiration date is not printed and there is a loss of useful lifetime, customers must rely on their external senses or sources of information to estimate the remaining useful life of a product.

3.1.4 Subtype Value

It refers to the economic value or utility given by the customer to each subtype in a given state. It can be the same or different for each subtype and, if the subtype state is dynamic, its value can change over time. For instance, in the fruit sector there is a clear dependence between price and quality [22]. Another factor that affects the price of fruit is the supply in relation to the demand. Due to the high fluctuations in the supply of crops, the agricultural sector usually perceives the price of fruit in the open market as random [23]. The supply and demand of vegetables and fruits is seasonal, and the market prices are not known in advance. For this reason, some products can be stored on the farm and sold when prices are higher [24]. Therefore, it can be said that the subtype value is uncertain and dependent on the subtype, the subtype state, the supply, the demand and the customers' and markets' preferences, which at the same time are greatly influenced by the environmental factors.

3.2 Resources

The resources used by the crop-based AFSCs include: fertilizers, pests, land, water, energy, machines and labour, storage and manufacturing facilities and transport means. The proper use of these resources largely impacts on ensuring sustainability: i.e. meeting current needs by using the provided resources in a way that the future generation's ability to satisfy their requirements is not compromised [25]. Kummu et al. [26] made a study about global food supply losses and how this affects the use of resources, such as freshwater, cropland, and fertilisers. They found that the food supply losses consume one quarter of these resources, and that half of them could be saved with proper planning, i.e., applying the minimum loss and waste percentages in each AFSCs step. Their findings highlight the large potential of efficient planning in AFSCs to save natural and scarce resources like freshwater.

3.3 Supply Chain Configuration

The SC configuration requires the definition at the Macro-Physical and Micro-Physical level. The first one shows how the network is configured and what material flows through it. Therefore, the elements to be modelled at the Macro-Physical level are: stages, nodes belonging to each stage, type of node (according to the type of activity to be done within the node: production operations, warehousing, selling points, or any combination of these) and arcs, which connect dyadic nodes and represent the flow of items from an origin node to a destination node [12]. Hoekstra and Romme [27] identify six basic types of designs that can be used to describe the relationships between actors: pipeline (one actor), chain (one supplier–one actor one customer), shared resource (several suppliers-one actor-several customers), converging (several suppliers-one customer), diverging (one supplier-several customers), and network (several suppliers-several customers) designs. At the Micro-Physical Level, the resources of each node are internally structured (e.g. facility layout and location, process type, etc.) and the composition of the arcs (e.g. transportation modes) that join two different nodes in the network.

4 Functional View

Activities that make up crop-based AFSCs have been identified by some authors with different levels of detail. This section first compiles these related works and then presents a proposal for an integrated crop-based AFSC decomposition. Along these lines, Ahumada and Villalobos [3] state that in the context of AFSC there are four main functional areas: production, harvest, storage, and distribution. Kusumastuti et al. [14] complete this definition for the fresh fruit supply chains detailing that there is an activity before production called planting/sowing. Others [4, 8] set that activities that comprise AFSC are: farming processing/production, testing, packaging, warehousing, transportation, distribution, and marketing.

Both Kusumastuti et al. [14] and Verdouw et al. [19] differentiate between the activities that compose supply chains of fresh products and processed products in fruits supply chains and in AFSC respectively. In the first case [14] the main activities of fresh products AFSC are cultivating, harvesting, pre-processing (activities such as washing and packing the product), storing and transporting. In the second case [19] the main activities of fruit supply chains with fresh products are: growing and harvesting, washing, sorting, grading, packaging, labelling, storage, distribution. Furthermore, Teimoury et al. [13] state that AFSC processed products have an additional activity after pre-processing called processing, where the intrinsic properties of the goods are changed and final products with a higher added value are created to satisfy customer demand. In contrast, Verdouw et al. [19] only recognize that in the case of fruit supply chains processed products, there is a processing activity after growing and harvesting that transforms fresh fruits into food products.

A less detailed proposal is the presented by Borodin et al. [28] where processing of products is also considered by dividing the AFSC activities in: production, storage, processing and distribution. Finally, Handayati et al. [29] have also taken into consideration the processed products of AFSC so they divide the AFSC in: cultivating, harvesting, post-harvest, transporting, processing, marketing, and distributing. Fuertes-Miquel et al. [25] include in their characterization of Brittany Horticulture SC the main activities of breeding new plant varieties, producing seeds and plants, test new varieties and other supporting activities such as test new technologies, training human labor, advising, supporting, coordinating and informing producers and regulate price. On the one hand, in order to extend the products shelf life, some activities where the product is not modified are done before distributing it to customers. Some of the post-harvesting operations are washing, sorting, grading, packaging and labelling [19]. On the other hand, for processed products, pre-processing consists in the preparation of vegetables and fruits to the processing activities. For example, washing a fruit before its processing.

By adapting and integrating the aforementioned authors, Fig. 3 shows a proposal for a decomposition of crop-based AFSC in detailed activities, which also distinguishes between supporting and main activities.

5   Organization View

According to Mintzberg [30], organisation structure comprises: (1) the establishment of tasks and (2) co-ordination of those tasks in order to realise objectives. Keuning [31] add (3) the definition of authorities and responsibilities of each task. Therefore, in order to define the Organization View, it is necessary to assign the activities of the Functional View to each actor (stakeholder) and define their relationships. Generally, the following actors can be distinguished in crop-based

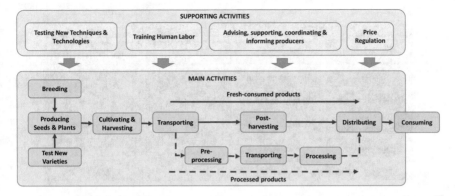

Fig. 3 Crop-based AFSC activities

AFSCs [25]: Research Centers & Institutes, Experimentation Centers, Training Centers, Growers, Farmers, Manufacture Processors, Cooperatives, Coordinator agents (e.g. Commerce Chamber), Traders (Exporters, Importers), Sorting and Packaging Stations, Consolidators, Auction Markets, Spot Markets, Dispatchers, Wholesalers, Retailers, Government Institutions and Customers. It is worth noting that crop-based AFSCs are traditionally very fragmented with a high number of small farmers and a predominant role of distributors. To properly manage these types of organizations, the level of integration becomes crucial.

6 Information and Uncertain View

To successfully execute any decision making-process on crop-based AFSCs it is necessary to get the required information. Decision-making is divided into risky and uncertain situations. In risky situations the decision maker knows both the alternative outcomes and the probability associated with each outcome. Under uncertainty, the decision maker does not know the probability of alternative outcomes [32]. In addition, when making a decision under uncertainty, the decision maker may or may not know the different outcomes that can occur [13].

Different classifications of AFSC uncertainties exist in the literature. Van der Vost [24] classifies generic SC uncertainties in a matrix with four uncertainty types (supply uncertainty, demand and distribution uncertainty, process uncertainty and planning and control uncertainty) affecting three aspects (quality, quantity and time aspects). Backus et al. [33] identify the following five uncertainties in the external environment: natural, technological, social, economic, and polítical factors. Grillo et al. [18] structures inherent lack of homogeneity in the product SC uncertainties in a matrix form with four uncertainty types such as: number of subtypes, subtype quantity, subtype state and subtype value, which that can appear in the different SC stages (supply, process and demand). On the other hand, Esteso et al. [34] propose

Fig. 4 Crop-based AFSCs uncertainty sources

four types of crop-based uncertainty: Product (shelf-life, deterioration rate, lack of homogeneity, food quality and food safety), Process (harvesting yield, supply lead time, resource needs, production), Market (demand, market prices) and Environment (weather, pests and diseases and regulations). More information on AFSCs under uncertainty can be consulted in Mundi et al. [35] and Grillo et al. [36].

In light of this, the proposed Conceptual Framework aims to classify the uncertainty in AFSCs (Fig. 4) by integrating the aforementioned uncertainties, but also by including others related to resources which, up to our knowledge, have not been identified in the literature yet: cost, availability (e.g., founds and labour) and quality of resources (e.g., land and water).

7 Decisional View

The above sources of uncertainties make it very difficult to match supply and demand in terms of customers' requirements of product varieties, qualities, quantities, times and prices. This imbalance, enhanced by the perishability aspect, frequently originates shortage and surplus situations accompanied with a great amount of waste. Poor management of these sources of uncertainties can have negative impact on crop-based AFSC efficiency and customer satisfaction. AFSC management involves a complex and integrated decision-making process that can be characterized by the following elements: Decision Structure (how the decisions are made at different levels and by different stakeholders?), Decisions (which decisions are made?) and Objectives (what are the pursued objectives when finding the value of the decisions to be implemented?).

7.1 Decision Structure

It is composed by the decision levels, the decision-makers at each level and their coordination/integration mechanisms. Three decision levels are usually considered: strategic, tactical and operational corresponding to long, medium and short time horizon, respectively. To achieve the temporal integration between levels, decisions made at higher levels should be respected by lower levels and/or disaggregated in order to be finally implemented on the physical system. Besides, at each temporal level, there can be one decision-maker (centralized decision-making) or several decision-makers (distributed decision-making). For this last case, it is necessary to define the coordination mechanisms among the existing decision-makers with the aim of achieving the spatial integration along the crop-based AFSCs.

7.2 Decisions

Several decisions are made at each decision level by specific decision-makers and at different AFSC stages. The most relevant decisions considered by different authors [3, 8, 15, 37] can be consulted in Fig. 5. It is important to highlight that depending on the Customer Order Decoupling Point location, decisions will be made based on demand forecasts (upstream the Customer Order Decoupling Point location) or on customer orders (downstream the Customer Order Decoupling Point location).

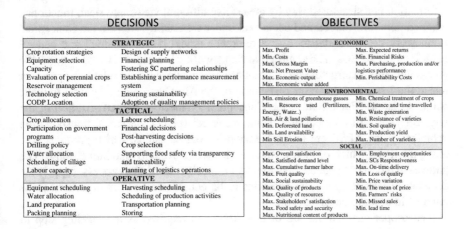

Fig. 5 Main crop-based AFSCs decisions and objectives

7.3   Objectives

The best value of the decisions will depend of the pursed objectives. Different types of SCs are defined based on their objectives. Farahani et al. [38] consider that a SC is *"sustainable"* when it considers in their objectives economic, environmental and social aspects. However, it is called *"green"* if it considers environmental and economic aspects, and is known as a *"lean"* when it considers only the economical aspect. Figure 5 shows the most common objectives of each type for crop-based AFSCs [25, 39–41]. Some of these objectives can be conflicting, being important to develop multi-objective or multi-criteria decision-making models.

8   Conclusions

The crop-based AFSCs operation has a significant impact on social, environment and economical aspects. However, the management of these supply chains becomes very complex due to their inherent products' characteristics, their sources of uncertainty and their fragmented organizational and decisional structure. Therefore, when making decisions at a specific temporal level (strategic, tactical or operational) for a specific AFSC process, it is necessary to gather the relevant aspects that will impact on the decisions to be made. Due to the inherent complexity of AFSC, this is not a trivial task.

In this paper, we propose a Conceptual Framework as a tool for gathering this information in a structured way by means five different views: physical, functional, organization, information and decisional under uncertainty. Indeed, the Conceptual Framework can be understood as a reference model (i.e. an abstract and general model) of any AFSCs. The Conceptual Framework when particularized to a specific AFSC will allow a structure description of its elements and their relationships under uncertainty, improving the understanding of the AFSC as a whole in order to make better decisions. For instance, the Conceptual Framework particularization of the physical view will make clearer which resources (plants, suppliers, transportation vehicles) of the AFSC are inside the scope of our decisions, which products flow along them and which of their characteristics are relevant to take into account. The particularization of the decisional view for a specific decisional process, support the decision-makers to guide the decisions to be considered and the objectives to be pursued in order to be sustainable.

The proper characterization of any AFSCs is required as a preceding step to develop reliable models to support its decision-making processes at the strategic, tactical and operational level under uncertainty. However, our proposal does not provide a direct relationship between the conceptual framework and the decisional models. For this, a more detailed methodology should be developed. The Conceptual Framework is a valuable tool for communication among decision-maker and/or researchers and AFSC stakeholders. Finally, it can be used to define the structural dimensions to analyse the literature on a specific decision-making process, identifying existing works and gaps for future research.

Acknowledgement Authors of this publication acknowledge the contribution of the Project 691249, RUC-APS "Enhancing and implementing Knowledge based ICT solutions within high Risk and Uncertain Conditions for Agriculture Production Systems" (www.ruc-aps.eu), funded by the European Union under their funding scheme H2020-MSCA-RISE-2015.

References

1. Taylor, D.H., Fearne, A.: Towards a framework for improvement in the management of demand in agri-food supply chains. Supply Chain Manage. **11**, 379–384 (2006)
2. Matopoulos, A., Vlachopoulou, M., Manthou, V., Manos, B.: A conceptual framework for supply chain collaboration: empirical evidence from the agri-food industry. Supply Chain Manage. **12**, 177–186 (2007)
3. Ahumada, O., Villalobos, J.R.: Application of planning models in the agri-food supply chain: a review. Eur. J. Oper. Res. **196**, 1–20 (2009)
4. Iakovou, E., Vlachos, D., Achillas, C., Anastasiadis, F.: A methodological framework for the design of green supply chains for the agrifood sector. Paper presented at the 2nd international conference on supply chains, Katerini, 5–7 Oct 2012
5. Manzini, R., Accorsi, R.: The new conceptual framework for food supply chain assessment. J. Food Eng. **115**, 251–263 (2013)
6. Shukla, M., Jharkharia, S.: Agri-fresh produce supply chain management: a state-of-the-art literature review. Int. J. Oper. Prod. Manage. **33**, 114–158 (2013)
7. Lemma, Y., Kitaw, D., Gatew, G.: Loss in perishable food supply chain: an optimization approach literature review. Int. J. Sci. Eng. Res. **5**, 302–311 (2014)
8. Tsolakis, N.K., Keramydas, C.A., Toka, A.K., Aidonis, D.A., Iakovou, E.T.: Agrifood supply chain management: a comprehensive hierarchical decision-making framework and a critical taxonomy. Biosyst. Eng. **120**, 47–64 (2014)
9. Van der Vorst, J.G., Da Silva, C.A., Trienekens, J.H.: Agro-industrial Supply Chain Management: Concepts and Applications. FAO (2007)
10. Hernandez, J., Mortimer, M., Patelli, E., Liu, S., Drummond, C., Kehr, E., Calabrese, N., Iannacone, R., Kacprzyk, J., Alemany, M.M.E., Gardner, D.: RUC-APS: enhancing and implementing knowledge based ICT solutions within high risk and uncertain conditions for agriculture production systems. In: 11th International Conference on Industrial Engineering and Industrial Management, Valencia, Spain (2017)
11. Miles, M.B., Huberman, A.M.: Qualitative Data Analysis: An Expanded Sourcebook. Sage Publications, Thousand Oaks (1994)
12. Alemany, M.M.E., Alarcón, F., Lario, F.C., Boj, J.J.: An application to support the temporal and spatial distributed decision-making process in supply chain collaborative planning. Comput. Ind. **62**, 519–540 (2011)
13. Teimoury, E., Nedaei, H., Ansari, S., Sabbaghi, M.: A multi-objective analysis for import quota policy making in a perishable fruit and vegetable supply chain: a system dynamics approach. Comput. Electron. Agric. **93**, 37–45 (2013)
14. Kusumastuti, R.D., van Donk, D.P., Teunter, R.: Crop-related harvesting and processing planning: a review. Int. J. Prod. Econ. **174**, 76–92 (2016)
15. Zhang, W., Wilhelm, W.E.: OR/MS decision support models for the specialty crops industry: a literature review. Ann. Oper. Res. **190**, 131–148 (2011)

16. Grillo, H., Alemany, M.M.E., Ortiz, A.: A review of mathematical models for supporting the order promising process under lack of homogeneity in product and other sources of uncertainty. Comput. Ind. Eng. **91**, 239–261 (2016)
17. Blanco, A.M., Masini, G., Petracci, N., Bandoni, J.A.: Operations management of a packaging plant in the fruit industry. J. Food Eng. **70**, 299–307 (2005)
18. Grillo, H., Alemany, M.M.E., Ortiz, A., Fuertes-Miquel, V.S.: Mathematical modelling of the order-promising process for fruit supply chains considering the perishability and subtypes of products. Appl. Math. Model. **49**, 255–278 (2017)
19. Verdouw, C.N., Beulens, A.J.M., Trienekens, J.H., Wolferta, J.: Process modelling in demand-driven supply chains: a reference model for the fruit industry. Comput. Electron. Agric. **73**, 174–187 (2010)
20. Amorim, P., Günther, H., Almada-Lobo, B.: Multi-objective integrated production and distribution planning of perishable products. Int. J. Prod. Econ. **138**, 89–101 (2012)
21. Nahmias, S.: Perishable inventory theory: a review. Oper. Res. **30**, 680–708 (1982)
22. Mowat, A., Collins, R.: Consumer behavior and fruit quality: supply chain management in an emerging industry. Supply Chain Manage. **5**, 45–54 (2000)
23. Kazaz, B., Webster, S.: The impact of yield-dependent trading costs on pricing and production planning under supply uncertainty. M&SOM Manuf. Serv. Oper. Manage. **13**, 404–417 (2011)
24. Van der Vorst, J.G.: Effective food supply chains: generating, modelling and evaluating supply chain scenarios (2000)
25. Fuertes-Miquel, V.S., Cuenca, L., Boza, A., Guyon, C., Alemany, M.M.E.: Conceptual framework for the characterization of vegetable breton supply chain sustainability in an uncertain context. In: 12th International Conference on Industrial Engineering and Industrial Management, XXII Congreso de Ingeniería de Organización, Girona, Spain, 12–13 July 2018
26. Kummu, M., de Moel, H., Porkka, M., Siebert, S., Varis, O., Ward, P.J.: Lost food, wasted resources: global food supply chain losses and their impacts on freshwater, cropland, and fertiliser use. Sci. Total Environ. **438**, 477–489 (2012)
27. Hoekstra, S., Romme, J.: Integral Logistic Structures: Developing Customer-Oriented Goods Flow. Industrial Press Inc., New York (1992)
28. Borodin, V., Bourtembourg, J., Hnaien, F., Labadie, N.: Handling uncertainty in agricultural supply chain management: a state of the art. Eur. J. Oper. Res. **254**, 348–359 (2016)
29. Handayati, Y., Simatupang, T.M., Perdana, T.: Agri-food supply chain coordination: the state-of-the-art and recent developments. Logist. Res. **8**, 1–15 (2015)
30. Mintzberg, H.: The Structuring of Organisations. Prentice-Hall, Upper Saddle River (1979)
31. Keuning, D.: Grondslagen Van Het Management. Stenfert Kroese, Houten (1995) (in Dutch)
32. Esteso, A., Alemany, M.M.E., Ortiz, A.: Conceptual framework for designing agri-food supply chains under uncertainty by mathematical programming models. Int. J. Prod. Res. (2018)
33. Backus, G.B.C., Eidman, V.R., Dijkhuizen, A.A.: Farm decision making under risk and uncertainty. Neth. J. Agr. Sci. **45**, 307–328 (1997)
34. Esteso, A., Alemany, M.M.E., Ortiz, A.: Conceptual framework for managing uncertainty in a collaborative agri-food supply chain context. In: IFIP Advances in Information and Communication Technology, vol. 506, pp. 715–724 (2017)
35. Mundi, I., Alemany, M.M.E., Poler, R., Fuertes-Miquel, V.S.: Review of mathematical models for production planning under uncertainty due to lack of homogeneity: proposal of a conceptual model. Int. J. Prod. Res. (2019)
36. Grillo, H., Alemany, M.M.E., Ortiz, A., De Baets, B.: Possibilistic compositions and state functions: application to the order promising process for perishables. Int. J. Prod. Res. (2019)
37. Soto-Silva, W.E., Nadal-Roig, E., González-Araya, M.C., Pla-Aragones, L.M.: Operational research models applied to the fresh fruit supply chain. Eur. J. Oper. Res. **251**, 345–355 (2016)
38. Farahani, R.Z., Rezapour, S., Drezner, T., Fallah, S.: Competitive supply chain network design: an overview of classifications, models, solution techniques and applications. Omega **45**, 92–118 (2014)
39. Banasik, A., Bloemhof-Ruwaard, J.M., Kanellopoulos, A., Claassen, G.D.H., van der Vorst, J.G.: Multi-criteria decision making approaches for green supply chains: a review. Flex. Serv. Manuf. J. 1–31 (2016)

40. Paam, P., Berretta, R., Heydar, M., Middleton, R.H., García-Flores, R., Juliano, P.: Planning models to optimize the agri-fresh food supply chain for loss minimization: a review. In: Reference Module in Food Science (2016)
41. Soysal, M., Bloemhof-Ruwaard, J.M., Meuwissen, M.P., van der Vorst, J.G.: A review on quantitative models for sustainable food logistics management. Int. J. Food Syst. Dyn. **3**, 136–155 (2012)

An Extension to Scenarios to Deal with Business Cases for the Decision-Making Processes in the Agribusiness Domain

L. Antonelli, G. Camilleri, C. Challiol, A. Fernandez, M. Hozikian,
R. Giandini, J. Grigera, A. B. Lliteras, J. Martin, D. Torres, and P. Zarate

Abstract With the aim of pushing innovation through information and communication technology in the agri-business field, working closely with farmers is essential. It is especially important to systematically capture their knowledge in order to analyze, propose and design innovation artifacts (in terms of software applications). In this article, we use Scenarios to capture the knowledge of the experts that is elicited in

L. Antonelli (✉) · C. Challiol · A. Fernandez · M. Hozikian · R. Giandini · J. Grigera ·
A. B. Lliteras · J. Martin · D. Torres
LIFIA, Facultad de Informatica, Universidad Nacional de La Plata, 50 esq 120, La Plata, Buenos Aires, Argentina
e-mail: leandro.antonelli@lifia.info.unlp.edu.ar

C. Challiol
e-mail: cecilia.challiol@lifia.info.unlp.edu.ar

A. Fernandez
e-mail: alejandro.fernandez@lifia.info.unlp.edu.ar

M. Hozikian
e-mail: marian.hozikian@lifia.info.unlp.edu.ar

R. Giandini
e-mail: roxana.giandini@lifia.info.unlp.edu.ar

J. Grigera
e-mail: julian.grigera@lifia.info.unlp.edu.ar

A. B. Lliteras
e-mail: alejandra.lliteras@lifia.info.unlp.edu.ar

J. Martin
e-mail: jonathan.martin@lifia.info.unlp.edu.ar

D. Torres
e-mail: diego.torres@lifia.info.unlp.edu.ar

G. Camilleri
SMAC Group, IRIT, 118 Route de Narbonne, 31062 Toulouse Cedex 9, France
e-mail: guy.camilleri@irit.fr

C. Challiol
CONICET, Buenos Aires, Argentina

© The Editor(s) (if applicable) and The Author(s), under exclusive license
to Springer Nature Switzerland AG 2021
J. E. Hernández and J. Kacprzyk (eds.), *Agriculture Value Chain — Challenges and Trends in Academia and Industry*, Studies in Systems, Decision and Control 280,
https://doi.org/10.1007/978-3-030-51047-3_3

early meetings previous to the definition of requirements. At those early stages, there are many uncertainties, and we are particularly interested in decision support. Thus, we propose an extension of the Scenarios for dealing with uncertainties. Scenarios are described in natural language, and it is very important to have an unbiased vocabulary. We complement Scenarios with a specific glossary, the Language Extended Lexicon that is also extended to decision support. According to V-model life cycle, every stage has a testing related stage. Thus, we also propose a set of rules to derive tests from the Scenarios. Summing up, we propose (i) an extension to Scenarios and the Language Extended Lexicon templates, (ii) a set of rules to derive tests, and (iii) an application to support the proposed technique. We have applied the proposed approach in a couple of case studies and we are confident that the results are promising. Nevertheless, we need to perform a further exhaustive validation.

Keywords Scenarios · Uncertainties · Decision support · Agri-business · LEL

1 Introduction

Agricultural processes are complex by nature because they rely on unpredictable conditions as weather or market demand, as well as human decision on biased opinions and incomplete information. Many people participate in the processes, usually with different objectives, background, experience and level of studies. Thus, it is hard to obtain a complete and accurate understanding of the whole process. This is the motivation of the RUC-APS project [1], a European funding project dealing with Enhancing and implementing Knowledge based ICT solutions within high Risk and Uncertain Conditions for Agriculture Production Systems.

Requirements engineering is one of the most important stages in software development. Errors made at this stage can cost up to 200 times to repair if they are discovered when the software is delivered to the client. There are two approaches to elicit requirements: classic and agile. Use Cases are a widely used tool to describe software requirements in a classical approach. There are different templates with different level of detail to use according to the definition of the requirements. Nevertheless, when the definition of requirements is vague, agile methods use User Stories to discover the requirements in an iterative and incremental way while the software application is developed.

A. Fernandez · R. Giandini · J. Grigera · A. B. Lliteras · D. Torres
CICPBA, Buenos Aires, Argentina

D. Torres
Depto. CyT, Universidad Nacional de Quilmes, Bernal, Argentina

P. Zarate
ADRIA Group, IRIT, Université de Toulouse, 2 Rue du Doyen Gabriel Marty, 31042 Toulouse
Cedex 9, France
e-mail: zarate@irit.fr

In both cases, with classic and agile development, the client should have a clear idea of the role of the technology and a vision about the software artifact that he needs. Although the requirements of the software are mainly described during the requirements engineering phase, there are some early meetings previous to the requirements stage to discuss needs, wishes and expectation. In these meetings, the objectives and boundaries of the software application are defined. Following these definitions, the requirements engineering stage can be performed to analyze and describe requirements (either with Use Cases or User Stories for example).

It is important to have tools to capture the information of the early meetings. Moreover, when stakeholders are not aware of how technology can help, it is necessary to support them. In this case, the IT team needs to learn about the domain and make proposals about innovation. We have been participating in the RUC-APS project with the aim of providing innovation in information and communication technology (ICT) to agriculture. In this period we have learned that agriculture is a field with no much integration with ICT. And it is a field with many uncertainties.

Some uncertainties are related to decision that farmers have to take and once taken, it cannot be changed. For example, the conducting system of the plants is related to make plants go upward or go down (as a bush). It should be defined before planting because both conducting systems need different distance between the plants. After the decisions made, it cannot be changed. Another decision is the training system to use. If it is decided that the plants go upward, some string is needed to help the plant to go upward. Besides a string, others elements can be used. Each element has different advantages and disadvantages. Then, the pruning system establishes how to cut the plant in order to allow it to go upward or down. All these decisions are related among them. Nevertheless, it is quite impossible to evaluate the impact of one decision on another. So, the decision has to be made one by one in a progressive process.

Others uncertainties are related to everyday situations that should be evaluated to react in consequence. For example, the temperature monitoring of a greenhouse must be set in order to keep it constant. We have seen that both types of uncertainties are captured in early meetings. Specifically, we are interested in innovation in Decision Support Systems, that is, to take a decision on the first type of uncertainties. Nevertheless, we also consider the second type of uncertainties.

In this article, we propose to capture the knowledge obtained from early meetings through Scenarios. We present an extension to the Scenarios to deal with uncertainty. We also propose to complement the description of the Scenarios with a particular glossary, the Language Extended Lexicon (LEL). An unbiased language is very important to understand the scenarios. We also propose an extension to the LEL to deal with uncertainty in order to capture business knowledge through Scenarios and complement them with LEL. According to the V-model development life cycle, the product obtained in each step (requirements, design, and codification) should be tested. Thus, we also propose a set of rules to automatically generate tests from Scenarios. We propose a set of rules to derive tests from Scenarios (only from Scenarios, not from LEL) in order to fulfill with the V-model. These tests derived from Scenarios should be used as input to design tests for the requirements.

The paper is organized in the following way. Section 2 introduces LEL and Task/Method models we use in this work. Section 3 details our contributions, i.e. a process to capture Scenarios and LEL, as well as the extensions to the Scenarios and the LEL to deal with uncertainty. Section 3 also presents the rules to derive tests from Scenarios using the task/method model as the specification. Finally, Sect. 4 discusses some conclusions.

2 Background

This section describes the base elements used in our approach. It describes the original template of Scenarios and LEL that we use to capture the knowledge about business cases. And it also describes the Task/Method model, the technique used to describe tests derived from the Scenarios through the set of rules proposed. The topics of this section are the base for our approach. In the next section, the original template of Scenarios and LEL are extended to deal with business cases for the decision-making processes in the agribusiness domain. In addition, in a further section, the Task/Method model is used to describe the proposed rules of our approach.

2.1 Scenarios

Scenarios describe interactions between users and a future system [2]. It is also used to understand the context of the application because they promote the communication when there is a great variety of experts [3].

Leite et al. [4] defines a scenario with the following attributes: (i) a title; (ii) a goal or aim to be reached through the execution of the episodes; (iii) a context that sets the starting point to reach the goal; (iv) the resources, relevant physical objects or information that must be available, (v) the actors, agents that perform the actions, and (vi) the set of episodes.

The following Scenario describes the activity of determining cultural labors for tomato production. It is important to mention that the scenario describes some task with uncertainty because many decisions have to be taken. Although people (the farmers and their leader) take the decision, some software application can be used to support the process. Moreover, the last episode of the scenario is related to describe the definitions arrived according to some standard. This task can also be supported by an application software since the application can receive the information about the decisions, organize and present according to the standard. Moreover, the task of writing the procedure can also be tested to assure if the report produced by a future application satisfies or not the standard.

Listing 1 Scenario about cultural labors

Title: Determine cultural labors
Goal: Decide the conducting system, the training system and the pruning policies that should be used
Context: Tomato production
Resources: conducting system techniques, training system techniques, pruning policies, standard to describe procedures, procedures for the cultural labors
Actors: farmers, leader
Episodes:
The farmers and their leader decide a conducting system to use
The farmers and their leader decide a training system to use
The leader establishes the pruning policies to apply
The leader writes a procedure according to the standard describing the conducting system, the training system, and the pruning policies

2.2 Language Extended Lexicon

The Language Extended Lexicon (LEL) is a glossary used to capture and describe the domain's language [5]. Terms (also called symbols) are classified into four types: Subject, Object, Verb, and State. Subjects represent an active element that performs actions. Objects are passive elements on which subjects perform actions. A verb is used to represent the actions. Finally, States represent situations in which subjects and objects can be located. A symbol is described by two attributes: (i) notion and (ii) behavioral responses. Notion describes the symbol denotation and explains its literal meaning. While Behavioral responses describe its connotation, that is, the effects and consequences of the relationship between the defined symbol and others symbols defined in the LEL [6].

The following examples describe one term of each symbol category. It is important to remark the example in Listing 4. The verb describes the activity of controlling the temperature of the greenhouse. This situation presents some uncertainty since the temperature can change and the farmer should act in consequence in order keep it within a specific range.

Listing 2 LEL subject: farmer

Subject: Farmer
Notion: Person that grows tomatoes in a shared lot with other farmers
Behavioral responses:
The farmer participates in the determination of the cultural labors
The farmer plant the tomatoes
The farmer grows the tomatoes
The farmer control the temperature of the greenhouse

Listing 3 LEL object: greenhouse

> **Object**: Greenhouse
> **Notion**: Place to grow tomatoes in a controlled environment
> **Behavioral responses**:
> The farmer plant tomatoes in a greenhouse
> The farmer control the temperature of the greenhouse

Listing 4 LEL verb: Control the temperature of the greenhouse

> **Verb**: Control the temperature of the greenhouse
> **Notion**: Action of monitoring the temperature in order to maintain it between certain range
> **Behavioral responses**:
> The farmers monitors the temperature
> The farmers aerate the greenhouse to descend the temperature
> The farmers close the windows of the greenhouse to increase the temperature

Listing 5 LEL state: Flowering

> **State**: Flowering
> **Notion**: Phenological state of the tomato, characterized by the appearance of leaves and flowers
> **Behavioral responses**:
> The tomato change to fruition state after the appearance of the first fruit

2.3 The Task/Method Model

The Task/Method model is a knowledge modeling paradigm that considers the reasoning as a task [7, 8]. Its main advantage is to have a declarative form to express knowledge which can be easily processed by tools such as execution engines [9].

A Task/Method model is composed of two sub-models: (i) the domain model and (ii) the reasoning model. The domain model describes the objects of the world that are used by the reasoning model. The reasoning model describes how a task can be performed. It uses two modeling primitives: (i) task and (ii) method.

A task is a transition between two world state families (an action) and is defined by the following attributes: (i) name that describes the task, (ii) par, a typified list of parameters handled by the task, (iii) objective, the goal state of the task, (iv) method, describes one way of performing a task.

A method is characterized by the following attributes: (i) heading, the identification of the task achieved by the method, (ii) preconditions which must be satisfied to be able to apply the method, (iii) effects, consequences of a successful application of the method, (iv) control, achievement order of the subtasks, and (v) subtasks. This paper focuses on some of the attributes described before. A full description of this modeling paradigm is performed by Camilleri et al. in [9, 10].

We use the Task/Method model in order to describe the tests that should be applied to the Scenarios. Then, the Task/Method model is processed by an execution engine to finally test the Scenarios. Listing 6 shows the example of a Task/Method model to test the scenario of determining cultural labor described in Listing 1. The Scenarios are descriptions in natural language, while the Task/Method model is a computer language, that is why some changes must be done to the names. For example, while "conducting system" is valid in a Scenario, in Task/Method should be translated as "conductingSystem". Moreover, Scenarios are too wide and abstract, not all the situation can be tested. In this example, we only define to test the method of writing the procedures in order to verify if the description of the procedures satisfies some standard. The procedures are definition from the decision of conducting system, training system and pruning policies performed previously. It is important to mention that Task/Method model is a hierarchical model, describing decompositions of tasks using other tasks. That is why the test is finally performed in method M14 (the fourth line in M1) of Listing 7.

Listing 6 Method determine cultural labor

Method: M1
Task: determineCulturalLabor
Control:
Decide (farmers, leader, conductingSystem);
Decide (farmers, leader, trainingSystem);
Establish (leader, pruningPolicies);
Write (leader, procedures, standard, conductingSystem, trainingSystem, pruning-Policies);

Listing 7 Method write procedures according to the standard

Method: M14
Task: Write (leader, procedures, standard, conductingSystem, trainingSystem, pruningPolicies)
Control:
message ("procedures do not conform to writing standards, stop");
stop;

3   Proposals

This section describes the contribution of this paper: (i) an extension to Scenarios and the Language Extended Lexicon templates, and (ii) a set of rules to derive tests from Scenarios.

We propose to capture the actual knowledge obtained from early meetings through Scenarios. Nevertheless, it is very hard to collect domain information in software development. Thus, the Scenarios extended by the proposed approach help to acquire hypothetical and unclear situations, so as to convert the Scenarios in real based and concrete Scenarios. In order to deal with the uncertainty, we extend the Scenarios with 4 more attributes: key decision, variables (frozen and contextual), identified risks and factors of uncertainty. These attributes capture relevant information to make a decision. For example "determine the conduction system" is a key decision. In Listing 1 it was defined as part of the episodes, but in fact, this element should be captured as a key decision. Then, the LEL is used to complement the Scenarios, and we also proposed an extension of the LEL to deal with uncertainty. For example, the temperature of the greenhouse is a variable that change according to the weather conditions (its context) and some actions must be done to maintain it in a specific range. This information should be captured by the LEL.

Scenarios and LEL are elements used to capture the business knowledge in early meetings. According to the V-model, this should be done on the top level of the V-model. Thus, in order to fulfill with the V-model, we also propose a set of rules to derive test from Scenarios. It is important to mention that the derivation is performed only from Scenarios, the LEL is not considered in this paper.

Thus, we provide techniques to use in the top level of the V-model (the business level): Scenarios (complemented with LEL) and business test (derived from Scenarios through a set of rules). These both elements can be used to produce the elements of the following level (requirements level). That is, Scenarios should be used to describe requirements, and business tests should be used to describe acceptance test.

The rest of the section is organized in the following way. First, a collaborative process to capture the knowledge to describe Scenarios is presented. Then, the extensions to the Scenarios and the LEL are described. Finally, the rules to derive tests described in Task/Method model from Scenarios are detailed.

3.1 A Collaborative and Iterative Process to Capture Knowledge

This section describes an iterative and incremental process to capture the knowledge from the stakeholders in an early stage of software definition. In that early stage, there are many uncertainties, that is why the most critical elements to capture are the potential decision and eventual variabilities that should be considered.

The process begins with the definition of the Scenarios by a multidisciplinary team. During the description of the Scenarios, it is common that specific terms of the domain appear, that are used by the experts of the domain but are unknown to the technical team members. These terms should be described in the LEL. Scenarios are used to capture multiple views and promote communication among stakeholders [3]. Moreover, a multidisciplinary team helps to obtain a broader perspective to define scenarios (Fig. 1).

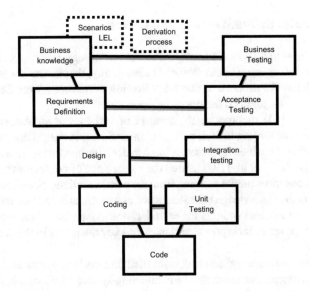

Fig. 1 Contribution located on the top of the V-model

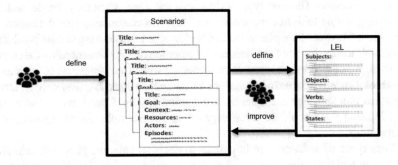

Fig. 2 Scenarios' definition process

It is important to state that the proposed process consists of describing mainly Scenarios, and describe terms in the LEL only if necessary. Traditionally, LEL is defined completely first and then Scenarios are described [4] (Fig. 2).

3.2 Extensions to Scenarios and LEL

The background section describes the original proposal to describe Scenarios and LEL. In order to support the capture of the knowledge in a very early stage of software definition, with many uncertainties and lack of precision, this section describe the attributes added to Scenarios and LEL.

3.2.1 Scenarios Extended to Deal with Uncertainty

In early meetings to discuss the incorporation of technology, it is important to add to the Scenarios information about factors of uncertainty, that is, the doubts that must be clarified later. We propose to add the following attributes: (i) key decisions, (ii) variables, (iii) risks, and (vi) factors of uncertainty.

Lupetti et al. [11] incorporate the concept of variables at an abstract level and categorize them according to their participation in Scenario design decisions. On the other hand, when a business process problem in decision support system is proposed, the work of De Maio et al. [12] stresses the importance of context-aware heterogeneous fuzzy consensus model learning from the past executions. So, there is a need to feed and maintain a knowledge base storing the associations between contextual variables, key decisions, and weights for each decision maker. Based on this reasoning, this article proposes to incorporate variables and key decisions in the description of the Scenarios.

Some uncertainties are related to decision that farmers have to take and once taken, it cannot be changed. For example, the conducting system mentioned before. Others uncertainties are related to everyday situations that should be evaluated to react in consequence. For example, the monitoring of the temperature of a greenhouse mentioned before. The first type of decisions are fixed when they are defined, and it makes possible to define more specific Scenarios considering that definition. For example the layout of the plantation can be discussed considering certain conducting system. While the second type, depends on situation that varies according the context.

Thus, we consider two types of variables: (i) frozen variables and (ii) contextual variables. Frozen variables are decisions that should be taken and they will not change from that moment. For example, the conduction system. Contextual variables refer to a decision that should be taken very often. For example, the monitoring of the temperature.

Both types of variables are important to decision support. People decide about the frozen variables, thus, they need tools to support the decision. While contextual variables are important to design the software application considering that it must react according to the variable.

In order to analyze the key decision, more information about the variables is needed. Both types of variable, mainly frozen variables, depends on risks and uncertainties. A risk is anything that could potentially impact your plan [13]. Thus, if a frozen variable for a decision is related to defining the growing space (indoor or outdoor), although the historical temperature can support the decision to grow outdoor, according to the geographical situation, there could exist the risk of unexpected low temperature.

The uncertainty concept is closely related to risks. A risk is an event that could potentially occur, thus, there is a measurable probability of occurrence. For example, there is a chance of 0.15 of low temperature during the harvest. But the factor of uncertainties is related to elements that there is no historical information, or cannot be estimated or predicted [14].

Listing 8 describes each proposed attribute with some information related to determine cultural labor scenario.

Listing 8 Extension to the Scenarios

Key decisions:
conduction system (How many main stems each plant will have?)
training system (How to support tomato plants off ground?)
pruning policies (When to start the pruning of stems and trusses to comply with the conduction system decisions? How often to prune?)
Variables
Frozen: soil type, seed type, geographic area, growing space (outdoor, greenhouse)
Contextual: diurnal temperature, nocturnal temperature, natural lighting
Identified risks: unexpected climate phenomena: frost, stronger winds, flood risk, low temperatures
Factors of uncertainty: plant disease, market demand, water pollution

Taking decisions (give values to frozen variables and deal with contextual variables) is related to analyzing and balance risk and uncertainties. From a classic point of view, it is necessary to analyze the risk involved in the key decisions, perform a quantitative and qualitative analysis of the uncertainties in order to make the decision [15, 16]. Of course, this decision could be made in different ways, but it is important to have documented the information to analyze in order to make the decision. That is the objective of adding these attributes to the Scenarios.

3.2.2 LEL Extended to Deal with Contextual Information

We propose to use Scenarios to capture uncertainty in order to take a decision. That is, experts and IT team analyze the scenarios and discuss alternatives to take a decision. Nevertheless, the uncertainty that the scenario represent could be a cause and effect relation, with different situations and different actions. In this case, is no need to take a decision, because all the alternatives should be considered by the application software. This, is another interpretation of decision making: a context-aware behavior, where relevant variables (context-features) take different values (representing specific situation) to trigger some decisions [12]. The context-aware behavior could present uncertainty when it is not possible to estimate or measure one or more variables' values. So, it is not possible to establish which decision should be triggered [17].

Litvak et al. [18] present an extension of LEL to provide more expressivity for Verbs (such as an action of an effect of), but, this is not enough to represent context-aware behaviors. Fortier et al. [19] model context-features and trigger decisions (handlers) as a first-class citizen due to the complexity involved in this kind of applications. Using the concepts defined in [19], this article proposes an extension to LEL that would help to define contextual information in a common vocabulary.

The LEL originally categorizes symbols in Subject, Object, Verbs, and States. This article proposes two new categories: Context-Feature and Contextual Decisions. The context-feature category allows representing each contextual feature associated to Subjects or Objects. Context-Features' values could be simple values (e.g. sensed data) or more complex elements [19]. Contextual Decision allows to represents for each situation (defined by a context-features' values) the list of triggered actions. The Given-then specification [20] can be used to describe behavioral responses for Contextual Decisions.

Let's consider the following examples. A Context-Feature "Temperature of greenhouse" which defines for some interval values trigger Contextual Decisions (Listing 9). And its Contextual Decision definition "Evaluate Temperature Decision" (Listing 10) that determines the action to be performed.

Listing 9 LEL Context-Feature: temperature of the greenhouse

Context-Feature: Temperature of the greenhouse
Notion: Range of temperature values measured in the greenhouse
Behavioral responses:
Value < 10 °C, Evaluate Temperature Decision
Value between 10 and 25 °C, No Action
Value > 25 °C, Evaluate Temperature Decision

Listing 10 LEL Context-decision: evaluate temperature decision

Context-decision: Evaluate temperature decision
Notion: When temperature is high or low, worker leader should be notified
Behavioral responses:
Given (Temperature of greenhouse <10 °C or Temperature of greenhouse >25 °C)
Then Notify to worker leader

3.3 Test Derivation from Scenarios

We propose five rules to derive test from the Scenarios [21]:

Rule 1 Tasks Identification: each verb in the Scenario's episodes is translated into a task in Task/Method model. Each Scenario title is also a task in Task/Method model. Listing 11 shows the example.

Listing 11

Title: Determine cultural labors → Task: DetermineCulturalLabors
Episodes
The farmers and their leader decide a conducting system to use → Task: Decide
The farmers and their leader decide a training system to use → Task: Decide

The leader establishes the pruning policies to apply $\rightarrow$ Task: Establish
The leader writes a procedure according to the standard describing the conducting
system, the training system, and the pruning policies $\rightarrow$ Task: Write

Rule 2 Task's Parameters Identification: each actor and resource used in the
episodes of a Scenario is translated into a parameter in Task/Method model. Listing
12 shows the example.

Listing 12

The farmers and their leader decide a conducting system to use $\rightarrow$ Task: Decide
(farmers, leader, conductingSystem)
The farmers and their leader decide a training system to use $\rightarrow$ Task: Decide
(farmers, leader, trainingSystem)
The leader establishes the pruning policies to apply $\rightarrow$ Task: Establish (leader,
pruningPolicies)
The leader writes a procedure according to the standard describing the conducting
system, the training system, and the pruning policies $\rightarrow$ Task: Write (leader,
procedures, standard, conductingSystem, trainingSystem, pruningPolicies)

Rule 3 Episode's Method: the episodes part of a scenario is translated by a method
in Task/Method model. Listing 13 shows the example. Since the main Scenario is
translated into the method named M1, each episode of the Scenario is translated into
a method M1#.

Listing 13

The farmers and their leader decide a conducting system to use $\rightarrow$ Method: M11
The farmers and their leader decide a training system to use $\rightarrow$ Method: M12
The leader establishes the pruning policies to apply $\rightarrow$ Method: M13
The leader writes a procedure according to the standard describing the conducting
system, the training system, and the pruning policies $\rightarrow$ Method: M14

Rule 4 The Sequence of Tasks: the sequence of different lines in the episodes part
of a Scenario determines the sequence of tasks in the control part of a Task/Method
model method. The use of expressions like "then", "after", etc.... in the episodes of
a Scenario determines also a sequence of tasks in the method's control part. Listing
14 shows the example.

Rule 5 Test Case Method: We assume that each test case (Test cases part of scenario)
corresponds to an achievement status (succeed or fail) of the task. In a failure situation,
the scenario will stop. This stop case will be represented by a method for the next
task in which the precondition field corresponds to the test case failure. The example
was shown in Listing 7.

Listing 14

Episodes:

...
The leader establishes pruning policies
The leader writes a guide...
Or
... The leader establishes pruning policies, then the farmers' leader writes a guide...
Method: M1
Task: determineCulturalLabor
Control:
...
Establish (leader, pruningPolicies);
Write (leader,procedures,standard, conductingSystem, trainingSystem, pruning-Policies);

4 Conclusions

We have presented a proposal to use Scenarios at an early stage of software development, when there are many uncertainties and the software is not defined yet. Also, gathering knowledge from farmers can be difficult, since they don't always organize their ideas in a way that's useful for requirements, so the proposed extensions to Scenarios and LEL, as well as the rules to derive tests, help acquiring information from them.

We enriched the Scenarios with some attributes that capture critical information to help stakeholders to take a decision and perform a further requirements definition of a software application. We complement the Scenarios with a particular glossary, the Language Extended Lexicon that we also extended to deal with uncertainty. According to the V-Model, in which every software development phase has a related testing stage, we also propose a technique to derive test from the Scenarios. Thus, we are providing a technique to deal with uncertainty and decisions on the top level of the V-model life cycle.

We have also built software applications to manage all the information. A Media Wiki platform [22] is used as a repository of the Scenarios and LEL. A semantic media Wiki extension was also added to allow the semantic support and the creation of forms in order to make the CRUD operations (create, retrieve, update and deleted) in a more user-friendly way. Then, in order to provide support to the derivation of tests, the tool relies on a Natural Language Processor Framework [23] and on a task automation tool and administrator of configurations [24].

This work was motivated with the aim of providing decision support to the agribusiness field in the context of the RUC-APS project. Although we have done some preliminary validation of the proposed strategy, we are planning to develop some pilot project in order to conduct case studies and to obtain feedback for improvement and validation of the proposal.

Acknowledgement Authors of this publication acknowledge the contribution of the Project 691249, RUC-APS: Enhancing and implementing knowledge based ICT solutions within high Risk and Uncertain Conditions for Agriculture Production Systems (www.ruc-aps.eu), funded by the European Union under their funding scheme H2020-MSCA-RISE-2015.

References

1. Hernandez, J., Mortimer, M., Patelli, E., Liu, S., Drummond, C., Kehr, E., Calabrese, N., Iannacone, R., Kacprzyk, J., Alemany, M., Gardner, D.: RUC-APS: enhancing and implementing Knowledge based ICT solutions within high risk and uncertain conditions for agriculture production systems. In: 11th International Conference on Industrial Engineering and Industrial Management, Valencia, Spain (2017)
2. Potts, C.: Using schematic scenarios to understand user needs. In: Proceedings of the 1st Conference on Designing Interactive Systems: Processes, Practices, Methods, & Techniques. ACM (1995). https://doi.org/10.1145/225434.225462
3. Carroll, J.M.: Five reasons for scenario-based design. Interact. Comput. **13**(1), 43–60 (2000). https://doi.org/10.1016/S0953-5438(00)00023-0
4. Leite, J.CSd.P., Rossi, G., Balaguer, F., Maiorana, V., Kaplan, G., Hadad, G., Oliveros, A.: Enhancing a requirements baseline with scenarios. Requirements Eng. **2**(4), 184-198 (1997). https://doi.org/10.1109/ISRE.1997.566841
5. Leite, J.C.S.d.P., Franco, A.P.M.: A strategy for conceptual model acquisition. In: Requirements Engineering Conference, pp. 243–246. IEEE. https://doi.org/10.1109/ISRE.1993.324851 (1993)
6. Sarmiento, E., Leite, J.C.S.d.P., Almentero, E.: C&L: generating model based test cases from natural language requirements descriptions. In: IEEE 1st International Workshop on Requirements Engineering and Testing (RET). IEEE (2014). https://doi.org/10.1109/RET.2014.690 8677
7. Schreiber, G., Akkermans, H., Anjewierden, A., De Hoog, R., Shadbolt, N.R., Wielinga, B.: Knowledge Engineering and Management: The CommonKADS Methodology, vol. 99 (2000)
8. Trichet, F., Tchounikine, P.: DSTM: a framework to operationalise and refine a problem solving method modeled in terms of tasks and methods. Expert Syst. Appl. **16**(2), 105–120 (1999)
9. Camilleri, G., Soubie, J.L., Zalaket, J.: TMMT: tool supporting knowledge modelling. In: Knowledge-Based Intelligent Information and Engineering Systems, vol. 2773, pp. 45–52 (2003)
10. Camilleri, G., Soubie, J.L., Zaraté, P.: Critical situations for decision making: a support based on a modelling tool. Group Decis. Negot. **14**(2), 159–171 (2005)
11. Lupetti, M.L., Gao, J., Yao, Y., Mi, H.: A scenario-driven design method for Chinese children edutainment. In: Proceedings of the Fifth International Symposium of Chinese CHI, pp. 22–29 (2017). https://doi.org/10.1145/3080631.3080636. ISBN: 978-1-4503-5308-3
12. De Maio, C., Fenza, G., Loia, V., Orciuoli, F., Herrera-Viedma, E.: A framework for context-aware heterogeneous group decision making in business processes. Knowl. Based Syst. **102**(2016), 39–50 (2016)
13. Project Management Institute: Project Management Body of Knowledge (2017)
14. Khodakarami, V., Fenton, N.E., Neil, M.: Project scheduling: improved approach to incorporate uncertainty using Bayesian networks. Proj. Manage. J. **38**(2), 39–49 (2007)
15. Trutnevyte, E., Guivarch, C., Lempert, R., Strachan, N.: Reinvigorating the scenario technique to expand uncertainty consideration. Clim. Change **135**(3), 373–379 (2016)
16. Weaver, C.P., Lempert, R.J., Brown, C., Hall, J.A., Revell, D., Sarewitz, D.: Improving the contribution of climate model information to decision making: the value and demands of robust decision frameworks. WIREs Clim. Change **4**, 39–60 (2013)

17. Borodin, V., Bourtembourg, J., Hnaien, F., Labadie, N.: Handling uncertainty in agricultural supply chain management: a state of the art. Eur. J. Oper. Res. **254**(2), 348–359 (2016)
18. Litvak, C.S., Hadad, G.D.S., Doorn, J.H.: Nominalizations in requirements engineering natural language models. In: Encyclopedia of Information Science and Technology, 4th edn, pp. 5127–5135. IGI Global (2018)
19. Fortier, A., Rossi, G., Gordillo, S.E., Challiol, C.: Dealing with variability in context-aware mobile software. J. Syst. Softw. **83**(6), 915–936 (2010)
20. Wynne, M., Hellesoy, A., Tooke, S.: The Cucumber Book: Behavior-Driven Development for Testers and Developers. Pragmatic Bookshelf (2017)
21. Antonelli, L., Camilleri, G., Grigera, J., Hozikian, M., Sauvage, C., Zaraté, P.: A modelling approach to generating user acceptance tests. In: International Conference on Decision Support Systems Technologies (ICDSST 2018), Heraklion, Greece (2018)
22. Media Wiki (2018). https://www.mediawiki.org
23. Stanford Natural Language Processing (2018). https://nlp.stanford.edu
24. Ansible IT Automation (2018). https://www.ansible.com/

Functional Value in Breeding Integrated to the Vegetables Value Chain as Part of Decision Making

C. Jana, G. Saavedra, N. Calabrese, J. P. Martinez, S. Vargas, and V. Muena

Abstract Actually, there is a great demand for healthy food containing added functional value, because people, have begun to change their diet with increments in the consumption of foods rich in antioxidants compounds, vitamins, minerals, and others. Vegetables are one of the most important sources of bioactive compounds; therefore, developing new varieties with enhanced content in bioactive compounds is an increasingly important breeding objective. Breeding programs should be focused, not just in pest resistance or tolerance, yield, and so on, but in the added value of the nutraceutical compounds, giving to new breeds special characteristics for human health. There have been efforts in some vegetables to generate new breeds with increased bioactive compounds, antioxidant as added value for the consumer, some of them are already commercial but there is still a long way to go. This chapter

C. Jana (✉)
Centro Regional de Investigación Intihuasi, Instituto de Investigaciones Agropecuarias, Colina San Joaquin S/N°, La Serena, Chile
e-mail: cjana@inia.cl

G. Saavedra
Centro Regional de Investigación La Platina, Instituto de Investigaciones Agropecuarias, Santa Rosa, 11610 Santiago, Chile
e-mail: gsaavedra@inia.cl

N. Calabrese
Institute of Food Production Sciences CNR ISPA, Via Amendola, 122/O, 70126 Bari, BA, Italy
e-mail: nicola.calabrese@ispa.cnr.it

J. P. Martinez · V. Muena
Centro Regional de Investigación La Cruz, Instituto de Investigaciones Agropecuarias, Chorrillos N° 86, La Cruz, Chile
e-mail: jpmartinez@inia.cl

V. Muena
e-mail: vmuena@inia.cl

S. Vargas
Centro Regional de Investigación Remehue, Instituto de Investigaciones Agropecuarias, Ruta 5 Sur, kilómetro 8, Osorno, Chile
e-mail: svargas@inia.cl

© The Editor(s) (if applicable) and The Author(s), under exclusive license to Springer Nature Switzerland AG 2021
J. E. Hernández and J. Kacprzyk (eds.), *Agriculture Value Chain — Challenges and Trends in Academia and Industry*, Studies in Systems, Decision and Control 280, https://doi.org/10.1007/978-3-030-51047-3_4

51

analyse the goals and advances in vegetables breeding, conventional and with use of biotechnology and their integration to the value chain, specifically in three important crops: artichoke, tomato and lettuce.

Keywords Functional foods · Bioactive compounds · Artichoke · Tomato · Lettuce · Breeding

1 Introduction

The concept of healthy food was born in Japan in the 80s, when the Japanese health authorities realized that they could control the public health expenses of a population with higher life expectancies, through foods that improved health or reduced risk of contracting diseases, which were called FOSHU, "Food with Specific Health Uses" [38]. The beneficial effect of the FOSHU, is for its ingredients (prebiotics, probiotics, antioxidants, omega-3 fatty acids, folic acid, phytosterols, phytoestrogens, among others), natural or added, or be-cause those components have been removed from it, food that can have a detrimental effect on health, such as the removal of allergenic, irritant, hypercaloric components, among others [45]. Within these are the functional foods (FF) are foods to which one or more healthy component naturally or added without changing its characteristics and healthy foods.

On the other hand, nutraceuticals (NT) is a word created in 1990 in the United States. NT is a more complex concept, since it is not attributed to a food but to components of a food and they are not pharmaceutical products, since they do not have therapeutic action although they can have preventive properties. It is a category for health care and whose effect is sustainable on the scientifically proven benefits of some nutrients and/or certain components (bioactive compounds) of foods of mainly vegetable origin, although it is also identified as some of the animal origin [31].

Vegetables contain many bioactive compounds and represent a major source of antioxidants and other compounds that are beneficial to human health or in the reduction of chronic non-transmissible diseases (CNTD) such as: cancer, cardiovascular disease, diabetes and obesity [43]. Given the increased demand for eat plants with increased bioactive compounds, researchers and breeders must develop new knowledge and tools for an efficient breeding of the content in bioactive compounds in vegetables [6, 8]. Value chain is a tool that is used to analyze all activities in a fragmented way, which are carried out within a process. Plant breeders must consider the value chain when they are breeding, to create more high performing varieties, customer focused and adopted by smallholder farmers.

In this chapter we deal with some relevant issues related to bioactive compounds in vegetables and breeding to increase the content with some examples in artichoke, tomato and lettuces, for their importance as vegetables of high functional value.

2 Properties About Main Phytochemicals in Vegetables

The main bioactive compounds found in plants are detailed below:

- Vitamin C (ascorbate) is a vital micronutrient for humans. Lack of vitamin C hinders the activity of a variety of enzymes and can cause scurvy in humans. Participates in the production of collagen, ascorbic acid, it serves as a cofactor in several vital enzymatic reactions. It is strongly suggested that vitamin C could prevent cardiac, chronic inflammatory and neurodegenerative diseases [40].
- Vitamin E (tocopherol) can prevent cardiovascular events, neuro-degenerative disease, macular degeneration and cancer. Has shown efficacy as anti-inflammatory and immune boosting compound. It has also shown some efficacy in protecting against nonalcoholic hepatosteatosis. At a molecular level, vitamin E and some of its metabolites have shown capacity of regulating cell signaling and modulating gene transcription [5].
- Phenolics compounds. There are a lot of plant phenolic compounds and representing secondary metabolites synthesized by plants during normal development in response to stress conditions. In plants, phenolics may act as phytoalexins, antifeedants, attractants for pollinators, contributors to plant pigmentation, antioxidants, and protective agents against UV light, among others. In food, phenolics may contribute to the bitterness, astringency, colour, flavour, odour, and oxidative stability of food [29]. The interest in polyphenols as antioxidants is focused on flavonoids which form a large family of low molecular weight polyphenolic compounds, which occur naturally in plant tissues and include the flavonols, flavones, flavanones, catechins, anthocyanins, isoflavonoids, dyhydroflavonols [40].
- Carotenoids (CARs) are very widespread in nature. Although CARs are typically seen as colorful pigments in fruits and flowers, they also occur in animals such as crustaceans and birds. CARs are essential components of photosynthetic membranes in plants and algae, although they are usually not seen in green tissues due to masking by the chlorophylls. Animals are unable to synthesize CARs de novo, and therefore food's diet is the main source of these compounds. CARs are also important dietary antioxidants and the major dietary precursor of Vitamin A [44], representing one of the largest classes of natural pigments.
- Phytosterols are found in high amounts in broccoli, brussels sprouts, cauliflower, and spinach. They regulate the fluidity and permeability of the phospholipid bilayers of plant membranes. Certain phytosterols are precursors of brassinosteroids, plant hormones involved in cell division, embryonic development, fertility, and plant growth. Some sterols are provitamins upon skin exposure to UV radiation; they may give rise to calciferol, also known as vitamin D2, which is involved in the absorption of calcium and bone growth. Plant sterols possess, moreover, cholesterol-lowering properties and play a positive role by decreasing the incidence of cardiovascular diseases. Plant sterols have been hypothesized to have anticancer, antiatherosclerosis, anti-inflammation, and antioxidant activities [40].

- Glucosinolates A reduction in the prevalence of certain forms of cancer has been attributed to the anticarcinogenic properties of certain glucosinolates and their breakdown products. Glucosinolates act by activating enzymes involved in the detoxification of carcinogens and by providing protection against oxidative damage. Certain glucosinolates have been observed to inhibit enzymes involved in the metabolism of steroid hormones [40]. Recent reports provide large number of evidences about the chemo preventive role of quercetin on various types of cancers such as bladder, prostate, esophagus, and in stomach cancers as well as towards other deleterious degenerative diseases [15].
- Saponins are non-volatile compounds. There are a big number of paper about that consumption of saponins increased the protection against the risk of cancer, decreased the level of blood cholesterol and lowered blood glucose response. In addition, anti-inflammatory, hypocholesterolemic and immune stimulatory activities of saponins have also been reported [28].

Numerous plant species have shown to be beneficial for human health due to its known biological activities, which include free-radical scavenging, regulation of enzymatic activity, and modulation of several cell signaling pathways, and others [39]. In fact, many of them are being actively studied as potential treatments for various metabolic and cardiovascular diseases. For example, brassica vegetables have a great impact on health promoting bioactive compounds; glucosinolates, flavonoids, hidroxycinnamic acid, and vitamin C. Saponins and lipids seem to be of relevance for the antidiabetic effects [11]; carrots with vitamin A of high antioxidant capacity provide health benefits beyond maintenance of vitamin A status [40]. The multiple health benefits of eggplant, which include anti-oxidant, anti-diabetic, hypotensive, cardioprotective, and hepatoprotective effects are largely attributed to its phenolic content, in particular to chlorogenic acid (CGA) [36]. It is also known to exert selective anti-carcinogenic effects via induction of apoptosis in many human cancer cells, such as leukemia cells [46] and lung cancer cells. Other biological activities of CGA include its antiobesity effect with improvement of lipid metabolism [7]. Leafy vegetables such as spinach contain many bioactive compounds including proteins, peptides, phenolics, tannins, saponins, cyanogenic-glycosides, terpenoids, alkaloids, steroids and defensins [2]. Phytochemical screening of tree spinach leaves confirmed the presence of secondary metabolites that inhibited *Escherichia coli* and *Bacillus subtilis* growth.

2.1 Properties About Artichoke, Tomato and Lettuce

Several studies have highlighted the health promoting properties of artichoke polyphenol components, such as inhibition of low-density lypoprotein (LDL) oxidation, hypoglycaemic, liver protecting and choleretic activities. Moreover, the in vitro scavenging effect on the reactive oxygen species (ROS) was also evaluated, showing a higher antioxidant activity of artichoke extract respect to the single main phenolic compounds [34]. The heads have a high content of protein, fiber, vitamins and minerals while its nutraceutical value flows from its content of inulin and various polyphenols [9, 21, 24]. Because the contents are variable depending the content of fertilization on soils, studies carried out in different fertilization regimes show that on average under a balanced system, the contents of reduced sugars and polysaccharides (glucose, fructose, sucrose and inulin) are high in relation to other species and accumulate in the receptacle [24, 25].

The chemical composition of fresh artichoke is reported in the Table 1.

Main phenolic and flavonoids compounds in artichoke in Table 2.

The importance of tomato as a healthy food is mainly explained by the different health promoting properties. Tomato contains several CARs in their fruit, representing the main source of lycopene (LYC) in our diet. Antioxidant abilities of LYC and beta-carotene are likely the mechanism of action of tomatoes preventing diet related diseases, specifically quenching reactive oxygen species (ROS). Besides CARs, polyphenolic compounds also contribute both to the health value of tomato improving the attributes of sensorial quality. These compounds are also thought to have health-promoting properties, probably due to their high antioxidant capacity. The function/activity of CARs and LYC has been extensively supported throughout by studies that have demonstrated its in vitro ability, to induce human protective enzyme systems. Similarly, a large series of epidemiological studies have suggested

Table 1 Chemical composition of artichoke (in 100 g of fresh weight)

Compounds	Organ	Quantity
Water (g)	Heads	91.3
Proteins (g)	Heads	2.7
Fats (g)	Heads	0.2
Sugars (g)	Heads	4.9
Fiber (g)	Heads	5.5
Sodium (mg)	Heads	133
Potassium (mg)	Heads	376
Calcium (mg)	Heads	86
Magnesium (mg)	Heads	45
Phosphorus (mg)	Heads	67
Iron	Heads	1.0
Copper	Heads	0.24
Zinc	Heads	0.95
Vitamin C (mg)	Heads	12
Niacin (mg)	Heads	0.50
Riboflavin (mg)	Heads	0.10
Thiamin (mg)	Heads	0.06

Adapted [32]

Table 2 Phenolic and flavonoid compounds based in HPLC analysis in artichoke (mg kg^{-1} of DM sample dry weight)

Compounds	Organ	Quantity
1-caffeoylquinic acid	Leaf	58 ± 4
	Head	23 ± 2
Chlorogenic acid	Leaf	481 ± 11
	Head	213 ± 7
Caffeic acid	Leaf	172 ± 6
	Head	275 ± 8
1,5-dicaffeoylquinic acid	Leaf	310 ± 8
	Head	160 ± 4
Luteolin	Leaf	53 ± 1
	Head	58 ± 3
Apigenin	Leaf	13 ± 2
	Head	29 ± 2

Adapted [32]

protective effects of CARs and LYC not only related with cardiovascular disease, but also against cancer and other age-related diseases such as dementia. Vitamins C and E are further naturally occurring antioxidants found in tomatoes [30].

The contents of lycopene vary among different tomato cultivars. A study carried out with 9 cultivars evaluating the lycopene content, found significant differences among them, which ranged between 2.95 and 6.06 mg 100 g (Fig. 1).

Several flavonoids have been reported in different tomato varieties, most of them as flavonols. Flavonol content usually depends on tomato variety, size, and country of origin, since light is one of the major environmental factors in the synthesis of flavonols. Vitamins C, including both ascorbic and dehydroascorbic acid, represents an important in the protection of tomato against autoxidative damage that might increase with ripening due to enhanced respiration [26].

Lettuce is one of the most widely consumed vegetables worldwide, but its nutritional value has been underestimated. Lettuce is low in calories, fat and sodium. It is a good source of fiber, iron, folate, and vitamin C. Lettuce is also a good source of various other health-beneficial bioactive compounds. In vitro and in vivo studies have shown anti-inflammatory, cholesterol-lowering, and antidiabetic activities attributed to the bioactive compounds in lettuce [10]. Nutrient composition and bioactive compounds vary among lettuce types. Crisphead lettuce, is comparatively low in minerals, vitamins, and bioactive compounds. More nutritious lettuces are leaf type lettuce and romaine with folate content comparable to other rich leafy vegetable sources. Red pigmented lettuce contains higher phenolic compounds than green lettuce. Baby green romaine is especially high in vitamin C (Fig. 2). Additionally, subtypes of leaf lettuce, such as continental, red oak, and lollo rosso lettuces have been reported to provide high levels of vitamin C [19].

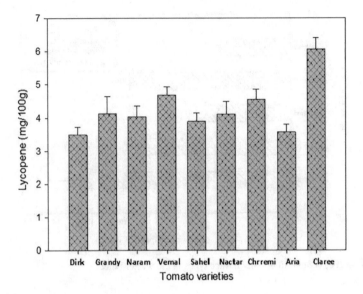

Fig. 1 Lycopene content of nine tomato varieties [16]

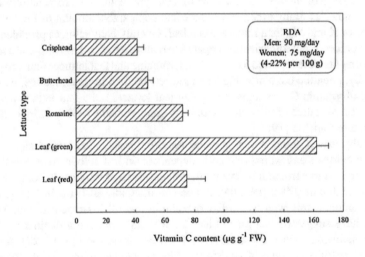

Fig. 2 Vitamin C content in popular lettuce types in the western diet

Studies generally have reported mineral content is higher in butterhead, romaine, and leaf lettuces than in crisphead (iceberg). Overall, mineral analysis show lettuce was a relatively good source of Fe (Fig. 3) and low in Na.

Lettuce is high in water content (95%) and low in calories and has nutritional benefits due to its contribution to dietary fibber, presence of several important dietary minerals, and various vitamins and bioactive compounds (carotenoids, and phenolic

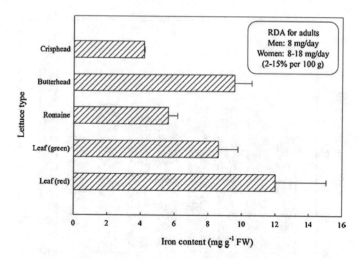

Fig. 3 Iron (Fe) content in popular lettuce types in the western diet

compounds). Among lettuce consumed, crisphead lettuce, known as iceberg, is the poorest source of dietary minerals, while leaf type lettuces were generally higher in these minerals. Baby green leaf lettuce had comparable amount of Fe, but lower contents of K and Mg than mature green leaf. Overall, baby lettuces provided equal or less number of phytonutrients compared to mature lettuces. Higher b-carotene and lutein contents were reported for butterhead, romaine and leaf lettuces than crisphead. Generally, phenolic content is higher in red lettuce than in green lettuce, while the content of vitamin C was highest in green leaf lettuce. Baby leaf lettuce may also provide a rich source of vitamin C with baby green romaine having a higher content than mature romaine [19].

In baby leaves, today a very popular food, the metabolites varied significantly between cultivars and were principally dependent on leaf colour. All red-leaf cultivars were rich in carotenoids, cyanidin, polyunsaturated fatty acids, total phenolic contents and antioxidant potential. Among carotenoids, all-E-lutein was found in highest amount, followed by all-E-violaxanthin and all-E-lactucaxanthin. Several studies have suggested that all red-leaf lettuce cultivars have a distinct profile of phytoconstituents, which can be used as a nutrient-dense food [3, 18, 27]. An evaluation of nutritional value of sixteen varieties of lettuces found that the flavonoid content and antioxidant activity in purple leafy lettuces were higher than that in green and red leafy lettuces, while the vitamin C and E contents in red leafy lettuces were higher [12]. Table 3, shows the content of bioactive molecules in purple and green lettuces.

Table 3 Content of several bioactive molecules in purple and green lettuce (mg/kg D.W)

Molecules	Purple lettuce	Green lettuce
Esculin	133.5 ± 13.9	50.7 ± 2.3
Chlorogenic acid	695.7 ± 56.9	137.9 ± 11.5
Aesculetin	148.9 ± 10.3	283.3 ± 8.5
Cynaroside	39.9 ± 3.3	222.7 ± 2.2
Lactucopicrin	93.5 ± 3.1	414.1 ± 22.3

[12]

2.2 Value Chain in Breeding

The characteristics to be breeding in a horticultural species are determined by different agents that make up the value chain. The demands of the consumer are those that have the greatest influence when selecting the characteristic to be breeding, since the vegetable producer is subject to market demand. The next agent in importance in the chain is the farmer and then, the rest of characteristic (Fig. 4).Knowing the steps and their risk and uncertainty in the value chain, allows take decisions into agricultural life cycle [13].

3 Breeding and Functional Value

Today we are speaking about "biofortification", this is an agricultural process that increases the uptake and accumulation of mineral nutrients in agricultural products through plant breeding, genetic engineering, or manipulation of agricultural practices

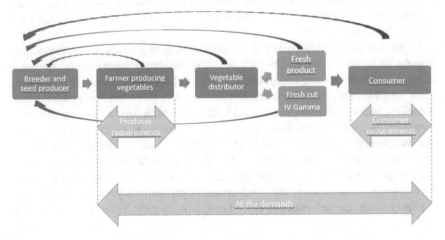

Fig. 4 Agents of the value chain in the horticultural production

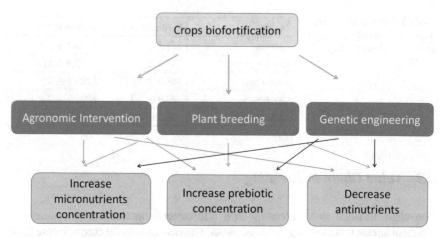

Fig. 5 Strategies to alleviate mineral deficiency. Adapted [14]

through agronomic approaches for improvement (Fig. 5).For example, nutritional value of lettuce can be enhanced by manipulating cultural practices. Water and N fertilization management as the more effective to manipulate the accumulation of plant-based phytochemicals which play an important role in human well-being [10].

Plant breeding is aimed at exploiting the genetic potential of plants for benefit of humans. Traditionally, the objectives of plant breeding programs have been aimed at improving yield, disease resistance, uniformity and quality, and often this improvement has had a reduction effect on bioactive compounds, which explains why traditional varieties have a greater quantity of these components than the new varieties. For example, gene *rin*, which is present in many long shelf-life varieties of tomato, causes a reduction in the content in lycopene in the fruit [20]. Furthermore, modern varieties usually have a narrow genetic base [1] and in order to improve the bioactive properties breeders very frequently will have to turn to materials like traditional varieties and wild relatives.

Given that there are important differences among vegetables in the compounds responsible for the bioactive properties, breeding programmers are usually directed to increase the levels of those compounds or groups of compounds that are responsible of the most relevant properties for each vegetable crop, for example in artichoke with phenolics, in particular chlorogenic acid, in lettuce with carotenoids, in particular β-carotene and lutein, and anthocyanins and in tomato in carotenoids, in particular lycopene, phenolic and ascorbic acid. Today there are an increasing number of breeding programmes and scientific studies aimed at improving the content in bioactive compounds of vegetables, and the trend seems that will be continuing in the coming years. In this respect, the development of genomics is greatly contributing to improve marker assisted selection as well as to develop tools for an efficient breeding [35].

3.1 Breeding Strategies

Starting point of a value chain is plant breeding (Fig. 4). If there is a demand for products with a greater quantity for bioactive components, is breeding, the one stable tools. Although some bioactive properties of specific vegetable crops may by qualitative (i.e., presence/absence), in most cases the traits responsible of the bioactive properties are quantitative high environmental influence in the expression of this type of traits, thus strategies for breeding for bioactive properties are those used for quantitative traits. We can find conventional strategies based on phenotyping, marker assisted selection, and strategies derived from genetic transformation (not conventional).

Conventional strategies based on selection in genetically variable populations for the trait of interest and on hybridization and selection in segregating generations [1]. Application of these breeding methods to traits related to bioactive properties shows that for these traits it is possible to achieve important genetic advances. It is not the case of artichoke, clonal propagation species, breeding is usually through mass selection of better performance plants. Selecting plants for traits such as cynarine content is not easy, due to the high environmental influence in the expression. Factors such as temperature, water availability and nitrogen source affect the amount of cynarin in the plant. Cynarine show differences in leaves, among different accession and growing season [37].

A different case is breeding in tomato, an autogamous plant, sexually propagated, with perfect flowers and self-compatible. Most varieties, actually, are hybrids and they have interesting traits and resistance/tolerance to abiotic and biotic stresses. Some of them have been selected for high lycopene content; a trait that industry looks for and in fresh consumption is an added value for the people. The concentration of this phytochemical is stimulated or inhibited by abiotic stresses, for instance, temperatures above 25 °C in fruits start the lycopene degradation, being completely destroyed at 40 °C, UV-B light and drought stimulates the lycopene generation [4]. Interspecific crosses have been the source for genes increasing the lycopene and other carotenoids, specially crossing with wild relatives such as *Solanum lycopersicon* var. *cerasiformes* and *S. pimpinellifolium* [22]. Breeding tomatoes for high lycopene content use strategies like recurrent selection for specific characters and introgression from wild accessions, but it takes time and it is difficult to eliminate wild traits. Actually, there are hybrids developed with specific purpose to produce higher contents of lycopene for chemical industry.

Lettuce is a natural inbreeder due to the structure of the flower and absence of self-incompatibility genes. There are not breeding programs for increase in nutritional value or antioxidants, the goals are different mostly oriented toward pests and stress resistance, agronomical characters and adaptation to specific environments. However, lettuce contributes to daily diet with some antioxidant, also affected by environmental conditions. For instance, phenolic compounds increase with heat and high UV light, ascorbate and tocopherol concentration increase with high temperature and carotenoids, also, increase with temperature, salinity and UV light [17, 23, 33].

While the conventional plant breeding works with the phenotype, modern biotech-nology: transgenics, molecular markers and genomics, allow to work directly with the genotype. Gene action estimates together with relatively high values for heritability indicate that selection for enhanced characteristic will be efficient. Modern genomics and biotechnological strategies, such as QTL detection, candidate genes approach and genetic transformation, are powerful tools for identification of genomic regions and genes with a key role in accumulation of bioactive compound in vegetables.

Molecular and genomic tools have been a revolution in breeding for bioactive properties. Thanks to the new developments it has been possible to identify quanti-tative trait loci (QTL) as well as genes and allelic variants of these genes involved in the synthesis of compounds responsible for bioactive properties as well as molecular markers linked to them [20, 42] This makes feasible in vegetable crops the marker assisted selection for traits related to bioactive properties [36]. This is very impor-tant to answer to new consumer, producer and processor requirements in the different steps of the value chain.

When there is inadequate genetic variability in the population is possible use Mutation breeding, for example in beans mutants increases the bioavailability of Fe, Zn y P in edible tissue [41]. Breeding in content of bioactive compounds can also be achieved by means of genetic transformation, which allows important increases in a short period of time. Genetic transformation requires the introduction using different transformation techniques of one or several genes from different organisms in the genome of the target species in order to achieve transgenesis.

However, transgenic varieties are suffering from an important rejection at the social level and it seems difficult that they representing a realistic alternative for the development of commercially accepted varieties. Cisgenesis, which consists in the genetic transformation resulting only in the introduction of genes obtained from materials sexually compatible with the donor variety could be an alternative, but the better is the combination of conventional and modern strategies to development of a new generation of vegetable varieties with enhanced content in bioactive compound, in an integral breeding approach.

4 Conclusions

Today there is a great demand for healthy food containing antioxidants compounds, vitamins, minerals and vegetables, are one of the most important sources of that bioactive compounds, for example, artichoke have a high content of protein, fiber, vitamins, minerals, inulin and various polyphenols; lettuce is high in water content and low in calories and has nutritional benefits due to its contribution to dietary fibber, presence of several dietary minerals, and vitamins and bioactive compounds or tomato, representing the main source of lycopene in our diet.

In the value chain, plant breeding must be the first step because it depends directly on consumer demand. However, many times it is not considered in its real importance. New breeding priorities and strategies must be developed for different crops.

There have been efforts in some vegetables to generate new breeds with increased antioxidant as added value for the consumer, some of them are already commercial, but in other crops such as lettuce and artichoke no visible results have been reported so far. There are many possibilities for the development of new cultivars with improved bioactive properties and it is necessary to continue making efforts through plant breeding. It is really important for future human nutrition that a simple vegetable portion content as vitamins, minerals and antioxidants, as be possible.

Acknowledgement Authors of this publication acknowledge the contribution of the Project 691249, RUC-APS: Enhancing and implementing Knowledge based ICT solutions within high Risk and Uncertain Conditions for Agriculture Production Systems (www.ruc-aps.eu), funded by the European Union under their funding scheme H2020-MSCA-RISE-2015.

References

1. Acquaah, G.: Principles of Plant Genetics and Breeding. Wiley-Blackwell, Chichester, UK (2012)
2. Adeniran, O., Olajide, O., Igwemmar, N., Orishadipe, A.: Phytochemical constituents, antimicrobial and antioxidant potentials of tree spinach. J. Med. Plants Res. **7**(19), 1310–1316 (2013)
3. Asadpour, E., Ghorbani, A., Sadeghnia, H.R.: Water-soluble compounds of lettuce inhibit DNA damage and lipid peroxidation induced by glucose/serum deprivation in N2a cells. Acta Pol. Pharm. **71**(3), 409 (2014)
4. Ayyagori, K., Sidhya, P., Pandit, M.K.: Impact of climate change on vegetable cultivation—a review. IJAEB **7**, 145–155 (2014)
5. Azzi, A.: Many tocopherols, one vitamin E. Mol. Aspects Med. Breed. **61**, 92–103 (2018)
6. Cámara, M.: Calidad nutricional y salud. In: Llácer, G., Díez, M.J., Carrillo, J.M., Nuez, F. (eds.) Mejora genética de la calidad, pp. 43–65. Universitat Politècnica de València, Valencia, Spain (2006)
7. Cho, A.S., Jeon, S.M., Kim, M.J., Yeo, J., Seo, K.I., Choi, M.S., Lee, M.K.: Chlorogenic acid exhibits anti-obesity property and improves lipid metabolism in high-fat diet-induced-obese mice. Food Chem. Toxicol. **48**, 937–943 (2010)
8. Diamanti, J., Battino, M., Mezzetti, B.: Breeding for fruit nutritional and nutraceutical quality. In: Jenks, M.A., Bebeli, P.J. (eds.) Breeding for Fruit Quality, pp. 61–80. Wiley, Hoboken, USA (2011)
9. D'Antuono, I., Di Gioia, F., Linsalata, V., Rosskopf, E.N., Cardinali, A.: Impact on health of artichoke and cardoon bioactive compounds: content, bioaccessibility, bioavailability and bioactivity. In: Petropoulos, S.A., et al. (eds.) Phytochemicals in Vegetables: A Valuable Source of Bioactive Compounds, pp. 316–393. ISBN 978-1-68108-740-5 (2018)
10. Galieni, A., Di Mattia, C., De Gregorio, M., Speca, S., Mastrocola, D., Pisante, M., Stagnari, F.: Effects of nutrient deficiency and abiotic environmental stresses on yield, phenolic compounds and antiradical activity in lettuce (*Lactuca sativa* L.). Sci. Hortic. **187**, 3–101 (2015)
11. Habicht, S., Kind, V., Rudloff, S., Borsch, C., Mueller, A., Pallauf, J., Yang, R., Krawinkel, M.: Quantification of antidiabetic extracts and compounds in bitter gourd varieties. Food Chem. **126**, 172–176 (2011)
12. Han, Y., Zhao, C., He, X., Sheng, Y., Ma, T., Sun, Z., Liu, X., Liu, C., Fan, S., Xu, W., Huang, K.: Purple lettuce (*Lactuca sativa* L.) attenuates metabolic disorders in diet induced obesity. J. Funct. Foods **45**, 462–470 (2018)

13. Hernandez, J., Mortimer, M., Patelli, E., Liu, S., Drummond, C., Kehr, E., Calabrese, N., Iannacone, R., Kacprzyk, J., Alemany, M., Gardner, D.: RUC-APS: enhancing and implementing knowledge based ICT solutions within high risk and uncertain conditions for agriculture production systems. In: 11th International Conference on Industrial Engineering and Industrial Management, Valencia, Spain (2017)

14. Huang, Y., Yuan, L., Yin, X.: Biofortification to struggle against iron deficiency. In: Yin, X., Yuan, L. (eds.) Phytoremediation and Biofortification Two Sides of One Coin, pp. 59–74. Springer, Dordrecht (2012)

15. Jan, A., Kamli, M., Murtaza, I., Singh, J., Ai, A., Haq, Q.: Dietary flavonoid quercetin and associated health benefits—an overview. Food Rev. Int. **26**, 302–317 (2010)

16. Khan, M., Javed, S., Butt, K., Nadeem, F., Yousaf, B., Javed, H.: Morphological and physico-biochemical characterization of various tomato cultivars in a simplified soilless media. Ann. Agric. Sci. **62**, 139–143 (2017)

17. Kim, H., Fonseca, J., Choi, J., Kubota, C., Kwon, D.: Salt in irrigation water affects the nutritional and visual properties of romaine lettuce (*Lactuca sativa* L.). J. Agric. Food Chem. **56**, 3772–3776 (2008)

18. Kim, D., Shang, X., Assefa, A., Keum, Y., Saini, R.: Metabolite profiling of green, green/red, and red lettuce cultivars: variation in health beneficial compounds and antioxidant potential. Food Res. Int. **105**, 361–370 (2018)

19. Kima, M.J., Moona, Y., Toub, J., Mouc, B., Waterlanda, N.: Nutritional value, bioactive compounds and health benefits of lettuce (*Lactuca sativa* L.). J. Food Compos. Anal. **49**, 19–34 (2016)

20. Kinkade, M., Foolad, M.: Validation and fine mapping of lyc12.1, a QTL for increased tomato fruit lycopene content. Theor. Appl. Genet. **126**, 2163–2175 (2013)

21. Lattanzio, V., Comino, C., Moglia, A, Lanteri, S.: Bio-active compounds and their synthetic pathway. In: Portis, E., et al. (eds.) The Globe Artichoke Genome, Compendium of Plant Genomes, pp. 99–113. https://doi.org/10.1007/978-3-030-20012-1_5 (2019)

22. Leiva, M., Valcarcel, M., Martí, R., Roselló, S., Cebolla, J.: New opportunities for developing tomato varieties with enhanced carotenoid content. Sci. Agric. **73**, 512–519 (2016)

23. Li, Q., Kubota, C.: Effects of supplemental light quality on grown and phytochemicals of baby leaf lettuce. Environ. Exp. Bot. **67**, 59–64 (2009)

24. Lombardo, S., Pandino, G., Ierna, A., Mauromicale, G.: Variation of polyphenols in a germplasm collection of globe artichoke. Food Res. Int. **46**, 544–551 (2012)

25. Lombardo, S., Pandino, G., Mauromicale, G.: The nutraceutical response of two globe artichoke cultivars to contrasting NPK fertilizer regimes. Food Res. Int. **76**, 852–859 (2015)

26. Mavromatis, A., Athanasouli, V., Vellios, E., Khah, E., Georgiadou, E., Pavli, O., Arvanitoyannis, I.: Characterization of tomato landraces grown under organic conditions based on molecular marker analysis and determination of fruit quality parameters. J. Agric. Sci. **5**, 239–252 (2013)

27. Meyer, G., Battault, S., Meziat, C., Gayrard, S., Aarouf, J., Urban, L.: Effects of low and high polyphenols content lettuces consumption on high fat diet induced metabolic syndrome and endothelial dysfunction. Arch. Cardiovasc. Dis. Suppl. **7**(2), 208–209 (2015)

28. Mudryj, A., Yu, N., Aukema, H.: Nutritional and health benefits of pulses. Appl. Physiol. Nutr. Metab. **39**, 1–8 (2014)

29. Naczk, M., Shahidi, F.: Phenolics in cereals, fruits and vegetables: occurrence, extraction and analysis. J. Pharm. Biomed. Anal. **41**, 1523–1542 (2006)

30. Navarro-González, I., García-Alonso, J., Periago, M.: Bioactive compounds of tomato: cancer chemopreventive effects and influence on the transcriptome in hepatocytes. J. Funct. Foods **42**, 271–280 (2018)

31. Nicoletti, M.: Nutraceuticals and botanicals: overview and perspectives. Int. J. Food Sci. Nutr. **63**, 2–6 (2012)

32. Nouraei, S., Rahimmalek, M., Saeidi, G.: Variation in polyphenolic composition, antioxidants and physiological characteristics of globe artichoke (Cynara cardunculus var. scolymus Hayek L.) as affected by drought stress. Sci. Hortic. **233**, 378–385 (2018)

33. Oh, M., Carey, E., Rajashekar, C.: Environmental stresses induce health-promoting phyto-chemical in lettuce. Plant Physiol. Biochem. **47**, 578–583 (2009)
34. Pandino, G., Lombardo, S., Mauromicale, G.: Chemical and morphological characteristics of new clones and commercial varieties of globe artichoke (Cynara cardunculus var. scolymus). Plant Foods Hum. Nutr. **66**, 291–297 (2011)
35. Pérez-de-Castro, A., Vilanova, S., Cañizares, J., Pascual, L., Blanca, J.M., Díez, M., Prohens, J., Picó, B.: Application of genomic tools in plant. Curr. Genomics **13**(3), 179–195 (2012)
36. Plazas, M., Andújar, I., Vilanova, S., Hurtado, M., Gramazio, P., Herraíz, F., Prohens, J.: Breeding for chlorogenic acid content in eggplant: interest and prospects. Notulae Botanicae Horti Agrobotanici **41**(2), 26–35 (2013)
37. Saavedra, G., Pino, M., Blanco, C.: Globe artichoke virus free 'Argentina' first year evaluation on field. Acta Hortic. **1147**, 375–380 (2016)
38. Saito, M.: Role of FOSHU (Food for Specified Health Uses) for healthier life. Yakugaku Zasshi **127**, 407–416 (2007)
39. Sato, Y., Itagaki, S., Kurokawa, T., Ogura, J., Kobayashi, M., Hirano, T., Sugawara, M., Iseki, K.: In vitro and in vivo antioxidant properties of chlorogenic acid and caffeic acid. Int. J. Pharm. **403**, 136–138 (2011)
40. Shashirekha, M., Mallikarjuna, S.E., Rajarathnam, S.: Status of bioactive compounds in foods, with focus on fruits and vegetables. Crit. Rev. Food Sci. Nutr. **55**(10), 1324–1339 (2015)
41. Shetty, P.: Incorporating nutritional considerations when addressing food insecurity. Food Secur. **1**(4), 431–440 (2009)
42. Sotelo, T., Soengas, P., Velasco, P., Rodríguez, V., Cartea, M.: Identification of metabolic QTLs and candidate genes for glucosinolate synthesis in *Brassica oleracea* leaves, seeds and flower buds. PLoS ONE **9**, e91428 (2014)
43. Terry, L.: Health-Promoting Properties of Fruit and Vegetables. CABI, Wallingford, UK (2011)
44. Wurbs, D., Ruf, S., Bock, R.: Contained metabolic engineering in tomatoes by expression of carotenoid biosynthesis genes from the plastid genome. Plant J. **49**, 276–288 (2007)
45. Yamada, K., Sato-Mito, N., Nagata, J.: Health claim evidence requirements in Japan. J Nutr. **138**, 1192–1198 (2008)
46. Yang, J., Liu, C., Ma, Y., Weng, S., Tang, N., Wu, S., Ji, B., Ma, C., Ko, Y., Funayama, S., Kuo, C.: Chlorogenic acid induces apoptotic cell death in U937 leukemia cells through caspase- and mitochondria dependent pathways. In Vivo **26**, 971–978 (2012)

Climate Change Mitigation Using Breeding as a Tool in the Vegetables Value Chain

G. Saavedra, S. Elgueta, E. Kehr, and C. Jana

Abstract Agriculture is a climate dependent activity, then both is affected by climate change and contributes to climate change. There is a vicious cycle that makes agriculture both a victim (because of negative effects of Global Warming on food supply) and a perpetrator (one of the main causes of Climatic Change). Producing more food for a starving world, agriculture requires large inputs, such as high amounts of nitrogen-based fertilisers, which in turn releases nitrous oxide (N_2O) emissions, one of the Greenhouse gases. Carbon dioxide, an important Greenhouse gas, for instance, can be beneficious for some vegetable species. There are other impacts of climate change in vegetable production, not only in plant physiology. Plant and fruit quality are affected by elevated temperatures; affect pollination and fruit set. There are other impacts on vegetable production systems, for instance, extended growing season because of warmer springs and autumns, but also new zones will be incorporated to vegetable production, where before it was impossible to produce larger amounts of quality vegetable for temperature. Mitigation of climate change is not an easy task; it involves many actors all over the science spectrum and government policies, but mainly the people conscience. Breeders may do huge efforts to generate varieties tolerant to many stresses. A challenge is to generate varieties tolerant to high temperature stress and low input production, with high rusticity and able to respond to biotic and abiotic stresses caused by Climate Change.

G. Saavedra (✉) · E. Kehr
Vegetables Department, INIA-Carillanca, Km 10 Camino Cajón, Vilcún, Temuco, Chile
e-mail: gsaavedr@inia.cl

E. Kehr
e-mail: ekehr@inia.cl

S. Elgueta
Vegetables Department, INIA-La Platina, Santa Rosa 11610, Santiago, Chile
e-mail: sebastian.elgueta@inia.cl

C. Jana
Vegetables Department, INIA-Intihuasi, Colina San Joaquín s/n, La Serena, Chile
e-mail: cjana@inia.cl

© The Editor(s) (if applicable) and The Author(s), under exclusive license
to Springer Nature Switzerland AG 2021
J. E. Hernández and J. Kacprzyk (eds.), *Agriculture Value Chain — Challenges and Trends
in Academia and Industry*, Studies in Systems, Decision and Control 280,
https://doi.org/10.1007/978-3-030-51047-3_5

67

Keywords Climate change · Global warming · Vegetables · Production · Greenhouse gases

1 Introduction

Global Warming, also referred to as "Climate Change", is defined by Encyclopaedia Britannica as the phenomenon of increasing average air temperature near the surface of Earth over the past one or two Centuries. However, modern Global Warming is the result of an increase in magnitude of the so called "Greenhouse Effect", a warming of Earth's surface and lower atmosphere caused by the presence of water vapour, carbon dioxide (CO_2), methane (CH_4), nitrous oxide (N_2O), and other greenhouse gases.

The global effects of Climate Change have been described such as increases in average and extreme temperatures all over the world, presenting extreme weather events in some regions, and accompanied with ice melt in polos which has brought increases in sea level and ocean acidification as result. But these changes, also, have affected plants and animals because of the modification of their growing environments, so many species will be vulnerable in the future and could disappear from Earth in the short term. The vulnerability of any system to Climate Change is the degree to which these systems are susceptible and unable to survive with the adverse impacts of Climate Change [1].

Agriculture is a climate dependent activity, then both is affected by Climate Change and contributes to Climate Change. Agriculture provides food for a world population increasing every year, but Would it be possible to supply enough food for the population, if there is loss of arable lands, increases in drought periods and new pests appear every time? United Nations estimates that 7.5 billion people today could increase to 9.7 billion people by 2050, but the crops production will decrease in 50% over the next 35 years because of altered climatic conditions. This fact will occur because of the impact of drought and severe weather, due to lack of accumulate snowmelts, lowering of water tables, diminish of rainfall, variations on temperature, and so on. Therefore, there will be loss of arable lands that will produce failure in several crops and livestock shortage worldwide. Even though, there will be expansion of growing season in some areas, a great number of diversities will be loss and harmful pests in new environments will appear damaging the crops [2].

In the other hand, agriculture contributes to Global Warming, as well. The industrial agriculture, currently employed by the majority of the developed world, has a hugely negative impact on Global Warming, through a huge carbon print. It is characterised by intensive agriculture and monocultures, mainly aimed to feed animal husbandry. This sector generates the highest amount of emission on CO_2, only compared to the total sum of CO_2 emitted by all forms of transportation. In summary FAO [3] estimates that 21% of all CO_2 emissions were caused by agriculture and deforestation in the decades from 2000 to 2010.

Agriculture needs an incremental amount of space in order to feed an increasing world population. Therefore, deforestation and paving over green space for suburban expansion result in more surface warming and a decreasing CO_2 capture. These facts will cause less absorption of CO_2 and mitigation of anthropic emissions.

There is a vicious cycle that makes agriculture both a victim (because of negative effects of Global Warming on food supply) and a perpetrator (one of the main causes of Climatic Change). Producing more food for a starving world, agriculture requires large inputs, such as high amounts of nitrogen-based fertilisers, which in turn releases nitrous oxide (N_2O) emissions.

Intensive agriculture, because of high fertilisers inputs release nitrates to the soil and water bodies. Also, high concentration of other nutrients, such as phosphates and nitrates, in water bodies, indirectly cause impact in eutrophication. This fact promotes algae growth and depletes oxygen in the water, affecting severely aquatic life and water quality.

In addition, the damage caused by fossil fuels combustion is other added problem, and then it is necessary to diminish the use of this kind of fuel, using more friendly types of energies.

Agriculture needs to increase yields while reducing dependence on agrochemicals, to reduce food waste, and to reduce consumption of resource-intensive and greenhouse gas-intensive foods such as meat.

2 Global Warming and Vegetables Production

As it has been mentioned before, vegetable production is dependent of climatic conditions to grow. Vegetables are the main source of essential nutrients, such as vitamins and minerals for the human body, but also, they supply other compounds in very few amounts, such as antioxidants, called as well nutraceutic compounds, which have an important role in the human metabolism.

However, environmental stresses due to Climate Change are major threats to global food security for the next years. Essentially, abiotic stress factors, such as drought, temperature raising, soil salinity, tropospheric ozone and excess UV radiation are cause already of important agricultural losses [4], where vegetables play a key role. These stresses affect directly vegetables production in different ways, not only in yield, also in product quality and post-harvest, altering the steps in the vegetables value chain.

2.1 Effects of Global Warming in Vegetable Physiology and Production

Several reviews have compiled information about the Global Warming effect in vegetables production [5–7] and the impact on crop quality [8]. The main effects described are related to Greenhouse Gases and temperature rising.

Carbon dioxide, an important Greenhouse gas, for instance, can be beneficious for some vegetable species, such as those belonging to the C3 group, which can be stimulated on their photosynthesis. The C3 plants have limited photosynthesis rate by CO_2 at current atmospheric conditions, an increase in CO_2 concentration can improve the performance in most vegetables species. Examples of some common vegetables are showed in Table 1. Researches in carrot [9] and bean [10] showed that a high CO_2 concentration and transpiration produces a decrease in stomatal conductance, then increase photosynthesis. But, high CO_2 concentration, also, can produce partial stomatal closure and transpiration may be reduced provoking an increase in the vegetables leaf temperature [11]. Therefore, it might reduce water use and make some vegetables more vulnerable in hot spells [5].

However, C4 plants such as sweet corn; photosynthesis is not affected by CO_2 concentration [15].

The normal historic CO_2 concentration has been between 180 and 220 ppm, but actually it has reached 350–400 ppm; therefore, concentrations over 1000 ppm is not feasible on field growing vegetables, it can be made just under controlled conditions, so far.

In the other hand, dark respiration by night should not exceed day photosynthesis rate, because a high photosynthesis/respiration ratio favors good yields. The rapid declination of dark respiration in roots and leaves of tomatoes, lettuce, peppers, peas, and maize plants grown under air enriched in CO_2 at night [16], indicated that in future scenarios with elevated atmospheric CO_2, night respiration will be lower, and then photosynthesis/respiration ratio will increase, especially when daytime photosynthesis will be stimulated [16, 17].

Another factor affecting vegetables production is elevated air temperature, essentially growth and yield. As many plants, vegetable crops physiological, biochemical and metabolic activities are temperature dependent and they have species specific temperature requirements. But, in the future is expected an increase in temperature that will change the growing conditions for vegetables. Mainly, because elevated temperature affects photosynthesis and respiration, there are specific thresholds for

Table 1 Improved performance in some vegetable species increasing CO_2 concentration

Species	CO_2 concentration [ppm]	Effect	
Tomato, cucumber, lettuce	720	44% increase in yield	[12]
Celery, lettuce, Chinese cabbage	800–1000	50% increase in yield	[13]
Onion seedlings	700	Twice dry matter content	[14]

each species, but in general vegetables belonging to C3 type maximum photosynthesis are reached between 20 and 32 °C [18], and C4 type like sweet corn needs 34 °C [15]. However, the response involves growth stage and plant age, as well. Elevated temperatures, above threshold, increase respiration rates and promote photorespiration, while photosynthesis decreases affecting negatively the ratio of the balance photosynthesis/respiration. But plants have the ability to acclimate and maintain the respiration rate slightest when temperature increases to heat stress conditions [19].

There are other impacts of climate change in vegetable production, not only in plant physiology. Plant and fruit quality are affected by elevated temperatures; in tomato temperatures above 25 °C affect pollination and fruit set, but also in fruits start the lycopene degradation, being completely destroyed at 40 °C. This high temperature plus raised UV light radiation produce "Sunburn" in fruits, damaging presentation and quality, but the most important fact decreasing the nutraceutic value due to lack of beta-carotenes and lycopene. Also, there is an inhibition of fruit ripening, because the suppression of ripening related m-RNAs, then there is not continuous protein synthesis including ethylene production, lycopene accumulation and cell-wall dissolution [20]. In pepper, elevated temperatures affect the development of red color in ripen fruits, but also provoke flower and fruit drop, besides poor fruit set.

In asparagus, excessive heating is a cause for quicker head opening, purple discoloration and fiber presence that reduces quality, price and consumer acceptance. In onions, there was a linear relation between temperature raising and organic sulfur contents, an important flavor precursor of pungency [21]. Similarly, to tomato, cucurbits vegetables temperature enhances abscission of flower buds, flowers and young fruits, also reduce fruit production, mature fruit size and seeds per fruit. But reduce fruit quality, as well, delaying ripening and diminishing the sweetness because of increasing respiration that consumes the sugars accumulated during photosynthesis, being the balance photosynthesis/respiration negative.

Leaf vegetables, such as lettuce, exposed to high temperature and radiation produce a common disorder called Tipburn that occur in the heart related to high transpiration and unequal calcium allocation. Besides, excessive radiation may damage lettuce and cabbage leaves, becoming papery and resulting in sunburn. But high temperatures may provoke early bolting and bitter compounds [18], then the product is not marketable. Other vegetables species such as sweet corn present a problem with tip fill due to elevated temperatures because pollen is not viable over 32 °C, broccoli shows loose heads and hollow stem, onions with split bulbs, blossom-end rot and sunburn in tomato and pepper are some examples of disorders caused by temperature in vegetables.

2.2   Effects of Global Warming in Vegetable Production Systems

There are other impacts on vegetable production systems, for instance, extended growing season because of warmer springs and autumns, but also new zones will be incorporated to vegetable production, where before it was impossible to produce larger amounts of quality vegetable for temperature. In the other hand, these raised temperatures during winter will provoke insufficient winter chill and vernalization in several species that require a stage of winter dormancy and accumulate cold temperatures to induce the spring growth, case of asparagus, rhubarb and artichokes. Many species are damaged with temperature between 12 and 16 °C for flower induction, usually bi-annual species such as *Brassicaceae*, *Alliaceae* and *Asteraceae* need a vernalization period. But, in some species such as lettuce, cabbage or spinach vernalization is undesirable because cause flower induction and spoils the product with a strong bitter taste and tough texture.

Increased temperatures could show impact on insect pest populations influencing ecology and biology [6]. Warmer temperatures in temperate climates will results in more types and higher populations of insects; for example, species with short life cycles such as aphids could increase fecundity, earlier completion of life cycle, producing more generations per year than their usual rate [22]. Also, elevated temperatures will cause migration of insect species to higher altitudes, while in the tropics higher temperatures might adversely affect specific pest species [6]. Therefore, new pests will appear in temperate climate, damaging crops that before were not attacked or the attack will last longer because of the length of high temperature due to climate change. Several of these pests are vectors of many viral diseases of vegetables crops (tomato, cucurbits, legumes and so on) causing severe loss in yield of these crops. In the case of diseases, higher temperatures produce faster biological cycles in the airborne pathogens and raise the persistence due to reduction of frost [2]. But, produce as well, early appearance and an increase in number of insect vectors of viral diseases due to rise of temperature during winter and implies a removal of climate limiting factors for many pathogens in vegetables.

Drought and salinity are the most important side effects of global warming, due to the lack of rainfall. In vegetable production, drought conditions and soil salinity produce serious problems in germination of seeds. Drought increases the salt concentration in the soil and affects the reverse osmosis of loss water from plant cells, increasing the water loss and inhibiting several physiological and biochemical processes such as photosynthesis, respiration and thereby reducing productivity in most vegetables [23]. There are some vegetables species more susceptible to salt stress such as onion, but some moderate tolerance is found in cucumber, eggplant, pepper and tomato.

Global warming, generally, increases plant biomass, indicating enhanced terrestrial Carbon uptake via plant growth, but there are complexity and many challenges in seeking general patterns in the future warmer world.

3 Problem Solution

Mitigation of climate change is not an easy task; it involves many actors all over the science spectrum and government policies, but mainly the people conscience. Breeders may do huge efforts to generate low inputs plants or with more rusticity, or tolerant to pests and diseases, however, if farmers and industrials are not prone to cooperate diminishing CO_2 release and other Greenhouse gases, or decreasing the carbon print it will be not possible to mitigate the effects in the world environment. Therefore, as part of the vegetables value chain, breeding programs have an important function as a vehicle by which new forms of production, technologies, logistics, labour processes, and organizational relations are introduced [24] to take part of the new breeds.

In general, agricultural production, probably vegetables are the crops that use more chemical inputs with a huge carbon print, behind corn, wheat and rice that are staple foods and food security is absolutely necessary to feed the world population.

3.1 Breeding Vegetables

A breeding program takes at least 15 years to have results, for this reason the breeder plans the breeds for the next 20 years or more. In this case, where climate change has evolved faster than it was thought 30 years ago, breeders have a huge challenge.

Carbon dioxide in vegetables is not really a problem, so far. However, elevated temperature produces important changes in product quality. A challenge is to generate varieties tolerant to high temperature stress, essentially lettuce, cabbage and leaves vegetables, that temperature induce bolting. Selection of breeds tolerant to temperature stress is important, but to improve efficiency of fertilizers is more important. Vegetables are crops very intensive in fertilizers use; normally farmers use more fertilizers than the crop's needs. For this reason, it is necessary to improve the use of fertilizers by vegetable plants, adjusting the amount and dressing, using more efficient system such as drop irrigation and fertilization. But the main idea is to create new breeds more efficient using nutrients. For instance, Table 2 is presented a list of common vegetables and nutrients removal by the consume organ [25], just to make an idea the real needs. The dressings exceed largely the amount need by whole plants, because both low efficiencies use by plants and dressing. Besides, fertilizers have a huge carbon print because of their industrial production and transport.

Other important fact in breeding vegetables is to decrease the use of chemical pesticides, because as fertilizers they have a huge carbon print. The strategy is to generate new varieties with tolerance/resistance to pests and diseases. For example, rootstock for grafting containing horizontal resistance to soil borne diseases and pests, that means several genes participate generating resistance, if one fail, still there are more acting. Therefore, fewer pesticides are needed. The program breeding may use wild parents as source of new genes resistant to pests and diseases, but not

Table 2 Twelve common vegetables and their nutrient removal by consume organ [25]

Species	Nitrogen (N)	Phosphorus (P_2O_5)	Potassium (K_2O)
Asparagus	6.1	1.8	4.5
Broccoli	1.8	0.5	5.0
Cabbage	3.2	0.7	3.1
Carrots	1.5	0.8	3.1
Lettuce	2.2	0.9	4.1
Onion	2.3	1.2	2.2
Peas	9.1	2.1	4.5
Sweet pepper	1.8	0.6	2.5
Spinach	4.5	1.2	5.4
Sweet corn	3.8	1.3	2.5
Tomato	1.8	0.4	3.2
Zucchini	2.1	0.7	3.0

only for this reason, because it may contribute with rusticity; then new varieties may tolerate thermal and water stress without large losses in production.

3.2 New Crops

In the world there are approximately 350,000 plant species, just about 80,000 are edible by humans, but actually only about 150 species are actively cultivated and 30 produce 95% of human calories and proteins [26].

There are several wild species and wild parents of vegetable crops that can be used both as gene sources and directly domesticated for use as food. The advantage of these species lay on adaptation to stressed environments (drought, salinity, elevated temperature), so they can be domesticated by selection using different genetic strategies. For example, some weeds are known because they are edible, such as thistle or cardoon rich in cynarine, *Portulaca oleracea* rich in Omega 3, and others.

Other regional crops have been introduced in different countries, some exotic Asian vegetables, South American solanaceous fruits and European specialties. But still there are species in the wild and others underutilized or neglected. This kind of crops presents a better opportunity for food crop, they are locally well adapted and constitute part of the local culture, require low inputs; and contribute to high agricultural sustainability [26].

Unfortunately, genetic improvement of wild species requires long time, involve high risk and investment, therefore much efforts must have been done from all point of view to carry out research and breeding for new vegetables crops more sustainable for the world.

3.3 Other Strategies

Adaptation strategies to climate change includes abiotic facts such as securing water and nutrient supply while reducing the damages caused by extreme weather events, as well as adapting the gene-pool of cultivars to new growing conditions.

Altered temperature and precipitation rates are expected to cause weeds to emerge earlier while rising CO_2 will increase their growth rate [17], ultimately increasing the competition between both, weeds and vegetables [12]. Similarly, pests and diseases will disseminate faster, with more biological cycles per year and causing much damage. So, additional applications of pesticides may be required increasing the carbon print of the crops; then more sustainable alternatives might be breeding of new cultivars better adapted but the use of insects screen, mulch layers (weed control and soil moisture) are technical approaches that could be useful.

Implementing innovative adaptations could enable vegetable producers to mitigate the adverse effects of global warming, while profiting from higher annual yields due to extended growing season and elevated CO_2 as well as higher antioxidant capacity as a response to abiotic aggressions. Changes in crop quality have received less attention than yield losses, perhaps because they are more difficult to detect, but abiotic stresses provoke a decreasing length in post-harvest of vegetables and presentation on shelves. The strategy for this problem is to have control on field of abiotic factors as much as possible and harvest a quality product.

Breeding strategies are an important part of the Vegetables Value Chain, if the following definition is strictly applied as it describes the full range of activities which are required to bring a product or service from conception, through the different phases of production (involving a combination of physical transformation and the input of various producer services), delivery to final consumers, and final disposal after use [27]. This input–output structure involves both goods and services, as well as a range of supporting industries. In order to understand the entire chain, it is crucial to study the evolution of the industry, the trends that have shaped it, and its organization [28]. Breeders releasing new varieties can modify phases of the production processes in vegetables value chain. New improved breeds selected by their performance mitigating Global Warming through low inputs production or rusticity traits, such as low fertilizers inputs or resistance to biotic stresses, they will diminish the carbon print because of less chemicals dressing, transport, less soil movement avoiding soil CO_2 captured releasing.

The breeding strategies not only involves improving traits within the germplasm available, but the use of wild parents, domestication or innovative technologies implementation are necessary for successful reduction on greenhouse gases by vegetables production. Several strategical combinations can be achieved and all of them will take part in the transformation of the value chain in vegetables production.

4 Conclusions

Vegetables production is both affected and affect global warming. There is consequence in plant physiology and morphology that produces increases/decreases in yield and quality products. Breeding new varieties tolerant to elevated temperature is a way to mitigate the climate change producing where before it was not possible, but most important is to generate breeds that diminish the use of agrochemicals using gene sources from the wild or old varieties, decreasing the carbon print of the final product. These facts will produce changes in the vegetables value chain, because the breeding strategies will lead to better products using less inputs and, therefore several phases will be modified. Accordingly, consumer will eat healthy vegetables, produced in a sustainable system and environmentally friendly.

Acknowledgement Authors of this publication acknowledge the contribution of the Project 691249, RUC-APS: Enhancing and implementing Knowledge based ICT solutions within high Risk and Uncertain Conditions for Agriculture Production Systems [29] (www.ruc-aps.eu), funded by the European Union under their funding scheme H2020-MSCA-RISE-2015.

References

1. Schneider, S.H., Semenov, S., Patwardhan, A., Burton, I., Magadza, C.H.D., Oppenheimer, M., Pittock, A.B., Rahman, A., Smith, J.B., Suarez, A., Yamin, F.: Assessing key vulnerabilities and the risk from climate change. Climate change 2007: impacts, adaptation and vulnerability, pp. 779–810. In: Parry, et al. (eds.) Contribution of Working Group II to the Fourth Assessment Report of the Intergovernmental Panel on Climate Change. Cambridge University Press, Cambridge, UK (2007)
2. Boonekamp, P.M.: Are plant diseases too much ignored in the climate change debate? Eur. J. Plant Pathol. **133**, 291–294 (2012)
3. Tubiello, F.N., Salvatore, M., Cóndor Golec, R.D., Ferrara, A., Rossi, S., Biancalani, R., Federici, S., Jacobs, H., Flammini, A.: Agriculture, Forestry and Other Land Use Emissions by Sources and Removals by Sinks. 1990–2011 Analysis. FAO Statistics Division, Working Paper Series ESS/14-02, 75 p (2014)
4. Ashmore, M., Toet, S., Emberson, L.: Ozone—a significant threat to future worldfood production? New Phytol. **170**, 201–204 (2006)
5. Bisbis, M.B., Gruda, N., Blanke, M.: Potential impacts of climate change on vegetable production and product quality—a review. J. Cleaner Prod. **170**, 1602–1620 (2018)
6. Ayyogari, K., Sidhya, P., Pandit, M.K.: Impact of climate change on vegetable cultivation—a review. Int. J. Agric. Environ. Biotechnol. **7**, 145 (2014)
7. Lin, D., Xia, J., Wan, S.: Climate warming and biomass accumulation of terrestrial plants: a meta-analysis. New Phytol. **188**, 187–198 (2010)
8. Wang, Y., Frei, M.: Stressed food—the impact of abiotic environmental stresses on crop quality. Agric. Ecosyst. Environ. **141**, 271–286 (2011)
9. Kyei-Boahen, S., Astatkie, T., Lada, R., Gordon, R., Caldwell, C.: Gas exchange of carrot leaves in response to elevated CO_2 concentration. Photosynthetica **41**, 597–603 (2003)
10. Radoglou, M.K., Aphalo, P., Jarvis, P.G.: Response of photosynthesis stomatal conductance and water use efficiency to elevated CO_2 and nutrient supply in acclimated seedlings of *Phaseolus vulgaris* L. Ann. Bot. **70**, 257–264 (1992)

11. Da Matta, F.M., Grandis, A., Arenque, B.C., Buckeridge, M.S.: Impacts of climate changes on crop physiology and food quality. Food Res. Int. **46**, 1814–1823 (2009)
12. Korres, N.E., Norsworthy, J.K., Tehranchian, P., Gitsopoulos, T.K., Loka, D.A., Oosterhuis, D.M., Gealy, D.R., Moss, S.R., Burgos, N.R., Miller, M.R., Palhano, M.: Cultivars to face climate change effects on crops and weeds: a review. Agron. Sustain. Dev. **36**, 12 (2016)
13. Jin, C., Du, S., Wang, Y., Condon, J., Lin, X., Zhang, Y.: Carbon dioxide enrichment by composting in greenhouses and its effect on vegetable production. J. Plant Nutr. Soil Sci. **172**, 418–424 (2009)
14. Bettoni, M.M., Mogor, A.F., Pauletti, V., Goicoechea, N.: Growth and metabolism of onion seedlings as affected by the application of humic substances, mycorrhizal inoculation and elevated CO_2. Sci. Hortic. **180**, 227–235 (2014)
15. Ruiz-Vera, U.M., Siebers, M.H., Drag, D.B., Ort, D.R., Bernacchi, C.J.: Canopy warming caused photosynthetic acclimation and reduced seed yield in maize grown at ambient and elevated $[CO_2]$. Glob. Change Biol. **21**, 4237–4249 (2015)
16. Peet, M.M., Wolfe, D.W.: Crop system responses to climatic change: vegetable crops. In: Reddy, K.R., Hodge, H.F. (eds.) Climate Change and Global Crop Productivity, pp. 213–243. CAB International, Oxon, New York (2000)
17. Mattos, L.M., Moretti, C.I., Jan, S., Sargent, S.A., Lima, C.E.P.: Climate changes and potential impacts on quality of fruit and vegetable crops. In: Emerging Technologies and Management of Crop Stress Tolerance, vol. 1, pp. 467–486 (2014)
18. Wien, H.C.: The Physiology of Vegetable Crops. CAB International, Oxon, New York (1997)
19. Körner, C.: Significance of temperature in plant life. In: Morrison, J.I.L., Morecroft, M.D. (eds.) Plant Growth and Climate Change. Blackwel, Oxford, UK (2006)
20. Lurie, S., Handros, A., Fallik, E., Shapira, R.: Reversible inhibition of tomato fruit gene expression at high temperature. Plant Physiol. **110**, 1207–1214 (1996)
21. Coolong, T.W., Randle, W.M.: Temperature influences flavor intensity and quality in Granex 33 onion. J. Am. Soc. Hortic. Sci. **128**, 176–181 (2003)
22. FAO: Climate-related transboundary pests and diseases, technical background document from the expert consultation held on 25 to 27 February 2008. FAO, Rome. Downloaded from ftp://ftp.fao.org/docrep/fao/meting/013/ai785e.pdf (2008)
23. Pena, R., Hughes, J.: Improving vegetable productivity in a variable and changing climate. J. SAT Agric. Res. **4**, 1–22 (2007)
24. Trienekens, J.H.: Agricultural value chain in developing countries a framework for analysis. Int. Food Agribusiness Manag. Rev. **14**, 51–82 (2011)
25. Warncke, D., Dahl, J., Zandstra, B.: Nutrient recommendations for vegetable crops in Michigan. Extension Bulletin E2934. Michigan State University, 31 p (2004)
26. Janick, J.: New crops for the 21st century. In: Nösberger, J., Geiger, H.H., Struik, P.C. (eds.) Crop Science: Progress and Prospects, pp. 307–327. CABI Publishing, Wallingford, Oxon, UK (2001)
27. Kaplinsky, R., Morris, M.: A Handbook for Value Chain Research. IDRC (2012)
28. Gereffi, G., Fernández-Stark, K.: Global Value Chain Analysis: A Primer. Center on Globalization, Governance & Competitiveness (CGGC). Duke University (2011)
29. Hernández, J., Mortimer, M., Patelli, E., Liu, S., Drummond, C., Kehr, E., Calabrese, N., Iannacone, R., Kacprzyk, J., Alemany, M., Gardner, D.: RUC-APS: enhanced implementing knowledge based ICT solutions within high risk and uncertain conditions for agriculture production systems. In: 11th International Conference on Industrial Engineering and Industrial Management, Valencia, Spain (2017)

Pesticide Residues in Vegetable Products and Consumer's Risk in the Agri-food Value Chain

S. Elgueta, A. Correa, M. Valenzuela, J. E. Hernández, S. Liu, H. Lu, G. Saavedra, and E. Kehr

Abstract The demand for food is increasing worldwide due to global population growth and the improvement of living standards. The use of pesticides in modern agriculture is necessary for most crops to guarantee the food supply. It has been estimated that the worldwide food production without the use of pesticides could decrease by as much as 35–40%, while the cost of food would increase. Vegetables and fruits are an important source of essential vitamins, minerals and antioxidants. However, the improper and excessive use of pesticides in agriculture includes the

S. Elgueta (✉) · A. Correa · M. Valenzuela
Laboratory of Pesticide Residues and Environment, Instituto de Investigaciones
Agropecuarias, CRI La Platina, Av. Santa Rosa 11610, Santiago, Chile
e-mail: sebastian.elgueta@inia.cl

A. Correa
e-mail: arturo.correabriones@gmail.com

M. Valenzuela
e-mail: marcela.valenzuela@inia.cl

J. E. Hernández
Management School, University of Liverpool, Liverpool, UK
e-mail: J.E.Hernandez@liverpool.ac.uk

S. Liu
Mast House, University of Plymouth, Plymouth PL4 8AA, Devon, UK
e-mail: shaofeng.liu@plymouth.ac.uk

H. Lu
Plymouth Business School, University of Plymouth, Devon PL4 8AA, UK
e-mail: haiyanlu@stu.edu.cn

G. Saavedra
Horticultural Department, Instituto de Investigaciones Agropecuarias, CRI La Platina,
Av. Santa Rosa 11610, Santiago, Chile
e-mail: gsaavedr@inia.cl

E. Kehr
Horticultural Department, Instituto de Investigaciones Agropecuarias, CRI Carillanca,
Camino 10 Km Vilcun, Temuco, Chile
e-mail: ekehr@inia.cl

J. E. Hernández and J. Kacprzyk (eds.), *Agriculture Value Chain — Challenges and Trends
in Academia and Industry*, Studies in Systems, Decision and Control 280,
https://doi.org/10.1007/978-3-030-51047-3_6

increasing of health risks. Nevertheless, some products can be a source of pesticide residues and their ingestion represents a potential source of diseases and negative effects to human health. This chapter describes principles and methods for health risk assessment of pesticide residues in vegetable and fruits. The exposure assessment of pesticide residues is an essential method to quantify the chemical risks. In addition, the dietary exposure assessment, combine food consumption data with the pesticide concentrations on vegetable and fruits. The aim is to support the decision making process in the agri-food value chain to improve the food safety enforcement and reduce the impacts on human health. The assessment can be developed for the problem formulation and the interactions between risk assessor, risk manager, stakeholders and decision-makers. In addition, the assessment based on scientific evidences facilitates the decision-making process in order to determine the impacts of pesticide residues on human health and set up their maximum residue limits-MRLs.

Keywords Pesticide residues · Human health · Decision-making process

1   Introduction

Pesticides are widely used in the agriculture to control pest and disease and to guarantee sufficient quantity of food [1]. The pesticides are classified by their use such as insecticides, fungicides, herbicides, rodenticides, fumigants and insect repellents and by their chemical structure such as organophosphates, pyrethroids, carbamates and organochlorines [2].

In the conventional cultivation, pesticides are used during the vegetation period to increase their yield prior to production or as post-harvest treatment [3]. The differences in the application periods, farm mixers, sprayer conditions, exposure, type of pesticides and agricultural areas where pesticides are applied are key factors to determine the chemical risks [4].

The use pattern of pesticides includes the application rates and pre-harvest intervals in farms both derived from the efficacy trials resulting in the lowest effective rate. A pesticide residue is any substance, mixture or cocktail resulting from their use in the farm. Their active ingredients and metabolites usually can be found after harvest time on fresh products. The concentration on vegetable and fruits can be reduced during storage, transport, preparation, commercial processing and cooking by consumers [5].

Countries usually monitored pesticide residues to determine their judicial use and unexpected residues that can affect the public health [6]. The consumption of vegetables and fruits is the primary way in which consumers are exposed to pesticide residues. The consumption patterns are set according the population at national level. The dietary intakes of pesticide residues within acceptable levels should not exceed the statutory established by governments around the world [7]. To provide information about chemical food safety, the periodic monitoring programs are needed to identify the improper use of pesticides in agriculture considering aspects to consumers as their potential exposures, the health risk associated and the miti-gation process [8]. Decisions on the level of pesticide residues and the amount of residues consumed are part of evaluations and approval process for their national registration and commercialization.

2 Pesticide Residues on Vegetable-Food Products

Pesticides and their residues in food are evaluated for their potential health risks and impacts in the environment. Their registration, authorization, commercialization and regulations depend on scientific evidences that can demonstrate no negative effects [9]. To investigate the safety of consumers, the exposure needs to be assessed and compared with health safety limits and toxicological end-point values such as the acceptable daily intake (ADI) and the acute reference dose (ARfD). Therefore at national and international level a Maximum Residue Limits (MRLs) for pesticides should be established [10].

The MRLs indicate the proper application of pesticides according to good agri-cultural practices in farms (GAP) and are strong indicators of violations of these practices. The MRLs represent thee maximum concentration of pesticide residues that is legally permitted at each country. The MRLs are expressed as milligrams of residue per kilogram.

The surveillance level at national level focuses on the proper use of pesticides for their authorization and registration in compliance with MRLs. One of the mandatory area is related to evaluate the impact on public health, which is mandatory. Decision-makers and authorities recognize the chemical safety of vegetable and fruits depends on economic, political and social risks. The food safety agencies at national and regional levels need to identify and organize the chemical hazards and manage the risks in order to implement appropriate policies and regulations to reduce the potential impacts on human health [11].

3 Risk Analysis Method for Pesticide Residues in Vegetables and Fruits

The analysis of pesticide residues in vegetables and fruits is relevant for a proper assessment of human exposure. The options for decision-makers about how to eval-uate the earliest stages of decision-making process are important. There are three steps to identify the framework in the food production in order to evaluate chemical risk [7].

The decision-making processes to reduce the chemical food safety issues can be evaluate through risk analysis method (Fig. 1). The first step is to include the problem

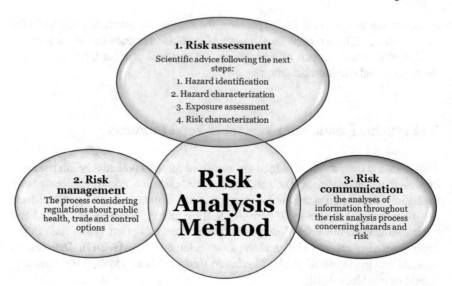

Fig. 1 Risk analysis method for the process of decision-making to reduce the chemical food safety issues

formulation and to scope the chemical safety conditions that can affect human health under risks detected to evaluate the possible risk management options.

The second step is planning and conducting the risk assessment to determine the levels of uncertainty and variability of the analysis. The third phase proposes to manage the risks considering the benefits of decisions and options to improve the system. The risk assessment should be transparent considering the uncertainties in the process and not be compromised for the involvement of decision–makers, scientific researchers and all stakeholders in the agri-food value chain [12].

3.1   Health Risk Assessment of Pesticide Residues in Vegetables and Fruits

The human exposure to low concentrations of pesticide residues through contaminated vegetables and fruits may lead to chronic toxicity. Chronic effects associated with chronic exposure to some pesticides include cancer, genetic mutations, endocrine disruptions and birth defects [13].

A full risk assessment is always carried on for the most vulnerable groups of society including children, elderly people and pregnant women, who may be especially vulnerable to be harmed by toxic effects [14]. The endpoints investigated include: genetic toxicity, acute oral toxicity, short-term toxicity, sub-chronic toxicity, long-term toxicity and carcinogenicity, reproductive toxicity, immunotoxicity and neurotoxicity. For the exposure assessment, the dietary intake of pesticide residues should be less than established accepted daily intake [15].

The interest in the applications of deterministic or probabilistic techniques for the estimation of human exposure to pesticide residues is growing internationally [10, 16, 17]. Determining exposure values based on pesticide residues and food consumption using deterministic or probabilistic approach have been used for health risk assessment [18]. The deterministic exposure assessment for acute and chronic dietary intake has been outlined in the World Health Organization guidelines for pesticide residues [19].

To evaluate health risk assessment can be used deterministic assumptions for major uncertainties there are pessimistic and optimistic [20]. The pessimistic model runs treat mayor uncertainties using assumptions expected to lead to over-estimation of exposure. The optimistic model runs treat major uncertainties using assumptions expected to lead to lower estimates of exposure [21].

For acute assessments, the resulting distribution may consider a lower estimate than the true distribution, since the exposure is unlikely to be lower. For chronic assessments less conservative than the basic pessimistic model, it cannot be guaranteed that the true exposure will not be underestimated and the model may be nearly as conservative as the pessimistic model [22].

3.2 Deterministic Assessments for Pesticide Residues

The preliminary step of risk assessment is to identify the regulatory process for pesticides residues at national level including their registration and authorization. The Maximum Residue Limits and the data obtained on national surveillance programs are necessary for the first step [23]. The deterministic models give an over-estimated exposure by assuming all time consumption of higher concentrations of pesticide residues [24].

To evaluate the chemical risks, the toxicological values as Acceptable Daily Intake, the Acute Reference Dose and consumption patterns are suitable to evaluate the health risks [25].

The Acceptable Daily Intake indicates the amount of pesticide residues that may be consumed every day for a lifetime without any appreciable risk. The Acute Reference Dose indicates the amount of pesticide residues ingested with one meal or in one day without any appreciable risks.

The predicting dietary intake of pesticide residue plays an important role in ensuring the chemical food safety for consumers at international level. There are different TIER approaches to predict pesticide residues intake.

The Theoretical Maximum Daily Intake (TMDI) uses MRLs to estimate the pesticide residues intake of consumers (Eq. 1). The assessment the average daily consumption for each vegetables and fruits and is an over-estimate of real concentration of pesticide residues intake.

The TMDI is compared to the ADI of the pesticide, for the average body weight of the population [19].

$$TMDI = \sum MRL \times F\,MRL \qquad (1)$$

where:

MRL: Maximum Residue Limits of vegetables and fruits
F: Per capita consumption of that vegetable-food products.

The International Estimated Daily Intake (IEDI) uses the correction factors and refines the intake estimations (Eq. 2). To estimate the current residue intake from vegetables food products, different factors should be used such as median pesticide residue levels from supervised trials, the pesticide residues in edible portions, the effects of processing and cooking on pesticide residue levels. If the ADI is exceeded by the IEADI, other sources as data from the government and industry will be necessary [19].

$$IEDI = \sum STMR \times E \times P \times F \qquad (2)$$

where:

STMR: Supervised trials median residue value of vegetables and fruits (mg/kg)
F: Consumption of vegetables and fruits
P: Processing factor for vegetables and fruits
E: Edible portion factor for vegetables and fruits.

The National Theoretical Maximum Daily Intake (NTMDI) can be used at national level to confirm the TMDI (Eq. 3). The use of MRLs of vegetables and fruits for estimating the dietary intake of pesticide residues can be used as a screening tool [19, 26].

$$NTMDI = \sum MRL \times F \qquad (3)$$

where:

MRL: Maximum Residue Limits of vegetables and fruits
F: National Consumption of vegetables and fruits per person.

3.3 *Probabilistic Assessments for Pesticide Residues*

The use of probabilistic assessments are complementary to deterministic. The methods introduce more realism representing the variation in consumption and allow quantification of uncertainties that influence the evaluation [27]. The software Monte

Carlo providing a range of risks throughout the population distributions, variability and uncertainties using the probabilistic assessment of potential exposures [28]. The estimation approach because of its simplicity and worldwide acceptance can be used as a screening method to identify pesticides possibly toxic for human health. Probabilistic models are more complexes and require more resources than deterministic approaches. It will therefore be logical to consider probabilistic models as an option for higher TIER assessment in those cases where deterministic methods are insufficient to reach a health risk management-decision [22]. With the use of probabilistic assessments, an important aspect is added to the risk assessment by consideration of the total consumption especially for the short-term dietary exposure. While using probabilistic assessments, the distribution of exposure amongst multiple individuals and the variability of food consumption, a more comprehensive assessment can be achieved.

3.4   Cumulative Risk Assessment of Pesticide Residues

In general, the consumers are exposed to more than one pesticide residue at the same time or within a short period. The combination of exposure to different pesticide residues may occur by the intake from a single vegetable and fruit with multiple pesticide residues from several vegetables and fruits with one or more pesticide residues.

All the interactions between pesticide residues include all forms of joint-action from dose or response addition [28]. The cumulative risk assessment can identify different pesticides residues as "cocktail effects" as function of their toxicities and can evaluate the combined effects of multiples pesticides [29]. Multiple routes of exposure (dietary, dermal, inhalation) can be important, especially for professional workers handling pesticides in farms [20].

The cumulating of toxicity of pesticides can be assessed using the Hazard Index or adjusted Hazard Index representing the sums of the ratios of individual pesticide exposure to their respective toxicological reference values. When the Hazard Index is less than 1, the combined risks from exposure to pesticide residues in the cumulative risk assessment is considered acceptable. All the assessment are sensitive to the choice of toxicological endpoint used to cumulate toxicity.

4   The Multi-criteria Decision Analysis in Chemical Food Safety

Chemical safety decision-makers face the challenge to consider consequences relating to more than one risk factor at time. They have a responsibility to promote the health of plants, animals and humans and the economy of each country. The multi-criteria decision analysis have been proposed as a method by operations research to compare complex issues considering aspect related toxicology, regulations, trade, food access, social and food security in the agri-food value chain.

At national level, the regulations and registration of pesticides should be establish considering different criteria that will be taken into consideration when reviewing a registration can be taken [30]. The criteria to choose the efficacy of pesticides and the risks to human health and the environment, including:

- The use of pesticides
- Occupational health hazard and risks to workers
- Hazards and risks to: public health, animal health and to the environment
- Quality of products
- Pesticide persistence and half life
- Pest resistance
- Proposed packaging and labels.

The different assessment for ranking the risks related to chemical safety in vegetables and fruits show a large variability in applications, emphasizing that each tool has its optimal purpose of use. More qualitative methods could be used when data are scare such as multi-criteria decision, risk matrix and expert judgment with an emphasis on inputs from experts [31]. It could simply be one decision maker or a group trying to foresee how different stakeholders might view a situation and they can prepare to address potential concerns [32].

The importance of protecting public health, the accepted practice is to prioritize food safety issues according to the risks to public health [33]. According to FAO, at national level several risks are relevant and are associated to public health and the environment, they may include:

- Food security and nutrition
- Economics earnings and impacts across stakeholders
- Impacts on food trade at local, regional and national levels
- Environmental concerns from food production and food industry
- Coherence with policy in areas such as rural development, animal health, environment and trade
- Consumers' perceptions and behavior
- Social, cultural and ethical considerations.

The multi-criteria can represent the opinion of stakeholders and can contribute to a balance between different stakeholders groups. The method is a structured process to establish priorities amongst diverse and competing policy options.

The purpose to evaluate multi-criteria is to enhance the transparency in the decision making process according the regulations at national level and to protect the public health and the environment. The method is relevant for food safety policies and interventions to create a national and integrated food-safety control system [34].

The first step is defineing the decision problem with inputs from stakeholders including the question to be addressed (Fig. 2). A list of interventions that can be

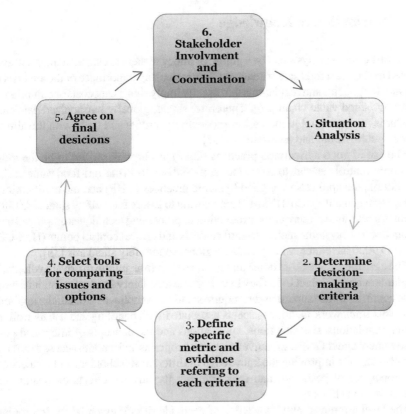

Fig. 2 Multi-criteria decision analysis overview [35]

applied to reduce the risks from pesticide residues will be necessary. The list consist in a scientific review with strong evidences related to the interventions. A key part of criteria is to measure the weight of evidences in order to quantify and incorporate the type of studies and their merits in terms of evidence [32]. To define specific metrics and evidences to each criteria is important to collect and analyze information related to public health at national level, pesticide toxicity, regulations and enforcement, trade, socio-economic issues. In addition to select tools for comparing issues and options, the risk management options using methods to determine uncertainty concerning the evidence and estimate it. Finally, the risk managers and stakeholders should validate the multi-criteria decision and agree on final decisions. The understanding is the first step toward compromise to establish transparent decisions that are relevant for stakeholders to gain consensus on the best approach to take. To determine decision-making criteria should be relevant, exhaustive, consistent, complete and clear.

5 Contribution to Knowledge

The multi-criteria analysis can be a guide for policy makers to establish interventions related to chemical food safety concerns. The increasing importance of the analysis is relevant for policies and regulations that usually imply significant costs of compliance in the agri-food value chain [36]. Therefore, single, globally harmonized pesticide standards are beneficial in increasing productivity, profits and trade with the aim to protect public health and environment [37].

The use of good agricultural practices (GAP) has been described to be the most important control measure to assure the safety and quality in the agri-food value chain. In addition, the application of good hygienic practices (GHP) and the certifications safety management systems (FSMS) are relevant to assure food safety standards [38]. There is a relevant scrutiny of the production or processing techniques employed and number of meta systems such as hazard analysis and critical control points (HACCP) and ISO 9000 have increasingly become global food safety standards [39].

The agri-food value chain come along with a shift from public to private voluntary standards such as GlobalGAP. They have become mandatory food safety standards in some countries due to supermarkets require standard compliance from their suppliers [40]. This framework identifies aspects associated with handling and use as well as current regulations, risk situations, applications and the concepts of integrated pest management under GAPs. Effective extension programs and smooth access to advice are necessary for improving the knowledge of different stakeholders and farmers on the proper use of pesticides and thus reducing the adverse effects on health and environment [41].

The level awareness and knowledge of pesticide risks is essential for improving safety in pesticides use. Awareness refers to consciousness of events or thoughts whereas knowledge has to do with information and skills acquired through experience or education. Knowledge influences on risk perception that in turn enhances safe pesticide practices [42]. The perception between farmers about awareness and knowledge is essential for improving safety in all aspects of pesticide handling. The knowledge is connected with understanding and sense of a concept, whereas awareness does not necessarily entail a deep understanding [43]. The lack of knowledge of farmers about the proper use of pesticide and the ignorance of potential health risks often results in increased pesticide exposure [44].

The lack of understanding of pesticide labels diminishes their effectiveness and safe use of pesticide. Some factors such as perceptions, lived experience, technical knowledge, training and education about pesticide can affect how users follow a label [45].

Usually when people get education and training related to food safety, they learn how to use all the job responsibilities that make up an integrated food safety system in the agri-food value chain [46]. Organized training programs should be held to provide adequate knowledge and skills for the safe use of pesticides among farm workers [47]. If the people receive a proper employee, training and understanding the food safety can be significantly improved [48]. To overcome deficiencies in the use and management of pesticides its relevant to create training courses on the management and use of pesticides in the agriculture [49].

6 Conclusions

The health risk assessment of pesticide residues is a complex and developing discipline in the center of pesticide approval process. There has been growing interest internationally in the application of probabilistic and deterministic assessments to estimate the exposure to pesticide residue in vegetables and fruits.

At the national level, decision-makers face the challenge of comparing and weighting the impact of different chemical safety issues. To formulate strong decisions to manage risks, they obtain and analyze relevant data to determine the impacts of pesticide residues obtained from the agriculture, elaborating a prioritization of relevant values.

The assessments will facilitate the decision making proccess of pesticide residues and their impacts on human health and allows for structured and transparent selection of risk asessement.

For effective engagement between agri-food stakeholders is essential to develop clear and transparent methodologies. A prioritization between common ideas for improving the coordination and consensus is necessary. Different factors such as social, economic and public health are the main issues for a better coordination between agri-food stakeholders. To validate the results the policy makers take into consideration these parameters and produce immediate, medium and long-term interventions depending on the resources of the country. After post-decision making process all, the regulations are more effective and sustainable. One of the main outcomes is the strengthening the food safety national surveillance programs that improve the management of different risks. Using the results of the process is expected to decrease the diseases and deaths produced by different food contaminations.

Acknowledgement Authors of this publication acknowledge the contribution of the Project 691249, RUC-APS: Enhancing and implementing Knowledge based on ICT solutions within high Risk and Uncertain Conditions for Agriculture Production Systems (www.ruc-aps.eu), funded by the European Union under their funding scheme H2020-MSCA-RISE-2015 and Project 502658-70 Ministry of Agriculture of Chile.

References

1. Cooper, J., Dobson, H.: The benefits of pesticides to mankind and the environment. Crop Prot. **26**, 1337–1348 (2007)
2. Oerke, E.: Crop losses to pests. J. Agric. Sci. **144**, 31–43 (2006)
3. Donkor, A., Osei-Fosu, P., Dubey, B., Kingsford-Adaboh, R., Ziwu, C., Asante, I.: Pesticide residues in fruits and vegetables in Ghana: a review. Environ. Sci. Pollut. Res. **23**, 18966–18987 (2016)
4. MacLachlan, D.J., Hamilton, D.: Estimation methods for maximum residue limits for pesticides. Regul. Toxicol. Pharmacol. **58**, 208–218 (2010)
5. Goh, K.T., Yew, F.S., Ong, K.H., Tan, I.K.: Acute organophosphorus food poisoning caused by contaminated green leafy vegetables. Arch. Environ. Health Int. J. **45**, 180–184 (1990)
6. Chen, C., Qian, Y., Chen, Q., Tao, C., Li, C., Li, Y.: Evaluation of pesticide residues in fruits and vegetables from Xiamen, China. Food Control **22**, 1114–1120 (2011)
7. Kapoor, U., Srivastava, M.K., Srivastava, A.K., Patel, D.K., Garg, V., Srivastava, L.P.: Analysis of imidacloprid residues in fruits, vegetables, cereals, fruit juices, and baby foods, and daily intake estimation in and around Lucknow, India. Environ. Toxicol. Chem. **32**, 723–727 (2013)
8. Abhilash, P.C., Singh, N.: Pesticide use and application: an Indian scenario. J. Hazard. Mater. **165**, 1–12 (2009)
9. Panuwet, P., Prapamontol, T., Chantara, S., Thavornyuthikarn, P., Montesano, M.A., White-head, R.D., Barr, D.B.: Concentrations of urinary pesticide metabolites in small-scale farmers in Chiang Mai Province, Thailand. Sci. Total Environ. **407**, 655–668 (2008)
10. Mojsak, P., Łozowicka, B., Kaczyński, P.: Estimating acute and chronic exposure of children and adults to chlorpyrifos in fruit and vegetables based on the new, lower toxicology data. Ecotoxicol. Environ. Saf. **159**, 182–189 (2018)
11. Schreinemachers, P., Tipraqsa, P.: Agricultural pesticides and land use intensification in high, middle and low-income countries. Food Policy **37**, 616–626 (2012)
12. Pitter, J.G., Jozwiak, Á., Martos, É., Kalo, Z., Voko, Z.: Next steps to evidence-based food safety risk analysis: opportunities for health technology assessment methodology implementation pathogen burden ranking for food safety risk prioritization. Stud. Agric. Econ. **117**, 155–161 (2015)
13. Parrón, T., Requena, M., Hernández, A.F., Alarcón, R.: Environmental exposure to pesticides and cancer risk in multiple human organ systems. Toxicol. Lett. **230**, 157–165 (2014)
14. Hines, R.N., Sargent, D., Autrup, H., Birnbaum, L.S., Brent, R.L., Doerrer, N.G., Hubal, E.A.C., Juber, D.R., Laurent, C., Luebke, R., Olejniczak, K., Portier, C., Slikker, W.: Approaches for assessing risks to sensitive populations: lessons learned from evaluating risks in the pediatric population. Toxicol. Sci. **113**, 4–26 (2009)
15. Hamilton, D., Ambrus, A., Dieterle, R., Felsot, A., Harris, C., Petersen, B., Rack, K., Wong, S.S., Gonalez, R., Tanaka, K., Earl, M., Roberts, G., Bhula, R.: Pesticide residues in food—acute dietary exposure. Pest Manag. Sci. **60**, 311–339 (2004)
16. Valcke, M., Bourgaul, M.H., Rochette, L., Normandin, L., Samuel, O., Belleville, D., Blanchet, C., Phaneuf, D.: Human health risk assessment on the consumption of fruits and vegetables containing residual pesticides: a cancer and non-cancer risk/benefit perspective. Environ. Int. **108**, 63–74 (2017)
17. Cequier, E., Sakhi, A.K., Haug, L.S., Thomsen, C.: Exposure to organophosphorus pesticides in Norwegian mothers and their children: diurnal variability in concentrations of their biomarkers and associations with food consumption. Sci. Total Environ. **590**, 655–662 (2017)
18. Jensen, B.H., Andersen, J.H., Petersen, A., Christensen, T.: Dietary exposure assessment of danish consumers to dithiocarbamate residues in food: a comparison of the deterministic and probabilistic approach. Food Add. Contam. Part A **25**, 714–721 (2008)
19. WHO.: Guidelines for Predicting Dietary Intake of Pesticide Residue. Document WHO/FSF/FOS/97.7 second revised ed. World Health Organization, Geneva (1997)

20. van der Voet, H., de Boer, W.J., Kruisselbrink, J.W., Goedhart, P.W., van der Heijden, G.W.A.M., Kennedy, M.C., Boon, P.E., van Klaveren, J.D.: The MCRA model for probabilistic single-compound and cumulative risk assessment of pesticides. Food Chem. Toxicol. **79**, 5–12 (2015)
21. Boon, P.E., van Donkersgoed, G., Christodoulou, D., Crépet, A., D'Addezio, L., Desvignes, V., Ericsson, B., Galimberti, F., Ioannou-Kakouri, E., Jensen, B.H., Rehurkova, I., Rety, J., Ruprich, J., Sand, S., Stephenson, C., Stromberg, A., Turrini, A., van der Voet, H., Ziegler, P., Hamey, P., van Klaveren, J.D.: Cumulative dietary exposure to a selected group of pesticides of the triazole group in different European countries according to the EFSA guidance on probabilistic modelling. Food Chem. Toxicol. **79**, 13–31 (2014)
22. EFSA Panel on Plant Protection Products and their residues (PPR): Guidance on the use of probabilistic methodology for modelling dietary exposure to pesticide residues. EFSA J. **10**, 2839 (2012)
23. Ambrus, Á., Yang, Y.Z.: Global harmonization of maximum residue limits for pesticides. J. Agric. Food Chem. **64**, 30–35 (2016)
24. Stephenson, C.L., Harris, C.A., Clarke, R.: An assessment of the acute dietary exposure to glyphosate using deterministic and probabilistic methods. Food Add. Contam Part A **35**, 258–272 (2018)
25. Lozowicka, B.: Health risk for children and adults consuming apples with pesticide residue. Sci. Total Environ. **502**, 184–198 (2015)
26. Maggioni, D.A., Signorini, M.L., Michlig, N., Repetti, M.R., Sigrist, M.E., Beldomenico, H.R.: Comprehensive estimate of the theoretical maximum daily intake of pesticide residues for chronic dietary risk assessment in Argentina. J. Environ. Sci. Health Part B Pest. Food Contam. Agric. Wastes **52**, 256–266 (2017)
27. Duan, Y., Guan, N., Li, P., Li, J., Luo, J.: Monitoring and dietary exposure assessment of pesticide residues in cowpea (*Vigna unguiculata L.* Walp) in Hainan, China. Food Control **59**, 250–255 (2016)
28. Jardim, A.N.O., Mello, D.C., Brito, A.P., van der Voet, H., Boon, P.E., Caldas, E.D.: Probabilistic dietary risk assessment of triazole and dithiocarbamate fungicides for the Brazilian population. Food Chem. Toxicol. **118**, 317–327 (2018)
29. Larsson, M.O., Nielsen, V.S., Bjerre, N., Laporte, F., Cedergreen, N.: Corrigendum to "refined assessment and perspectives on the cumulative risk resulting from the dietary exposure to pesticide residues in the Danish population." Chem. Toxicol. **113**, 345–346 (2018)
30. FAO.: Food and Agriculture Organization of the United Nations. International code of conduct on pesticide management. E-ISBN 978-92-5-108896-8 (2015)
31. van der Fels-Klerx, H.J., van Asselt, E.D., Raley, M., Poulsen, M., Korsgaard, H., Bredsdorff, L., Nauta, M., Flari, V., d´Agostino, M., Coles, D., Frewer, L.: Critical review of methodology and application of risk ranking for prioritisation of food and feed related issues, on the basis of the size of anticipated health impact. EFSA Supporting Publication, EN-710 (2015)
32. Fazil, A.M., Rajic, A., Sanchez, J., McEwen, S.A.: Choices, choices: the application of multi-criteria decision analysis to a food safety decision-making problem. J. Food Prot. **71**, 2323–2333 (2008)
33. FAO.: Food and Agriculture Organization of the United Nations. Food safety risk management, evidence-informed policies and decisions, considering multiple factors. Food Safety and Quality series. ISSN 2415-1173 (2017)
34. FAO.: Food and Agriculture Organization of the United Nations. Evidence-informed food safety policies and risk management decisions. FAO Technical meeting (2013)
35. FAO.: Testing multi-criteria approaches for food safety decision-making. Pilot Project Report (2014)
36. Ehrich, M., Mangelsdorf, A.: The role of private standards for manufactured food exports from developing countries. World Dev **101**, 16–27 (2018)
37. Handford, C.E., Elliott, C.T., Campbell, K.: A review of the global pesticide legislation and the scale of challenge in reaching the global harmonization of food safety standards. Integr. Environ. Assess. Manag. **11**, 525–536 (2015)

38. Boxstael, S., Van Habib, I., Jacxsens, L., Vocht, M., De Baert, L., Perre, E., Van De Rajkovic, A., Lopez-Galvez, F., Sampers, I., Spanoghe, P., Meulenaer, B., Uyttendaele: Food safety issues in fresh produce: bacterial pathogens, viruses and pesticide residues indicated as major concerns by stakeholders in the fresh produce chain. Food Control 32, 190–197 (2013)

39. Henson, S., Jaffee, S.: Understanding developing country strategic responses to the enhancement of food safety standards. World Econ. **31**, 548–568 (2008)

40. Handschuch, C., Wollni, M., Villalobos, P.: Adoption of food safety and quality standards among Chilean raspberry producers—do smallholders benefit? Food Policy **40**, 64–73 (2013)

41. Damalas, C.A., Khan, M.: Pesticide use in vegetable crops in Pakistan: insights through an ordered probit model. Crop Prot. **99**, 59–64 (2017)

42. Wang, J., Tao, J., Yang, C., Chu, M., Lam, H.: A general framework incorporating knowledge, risk perception and practices to eliminate pesticide residues in food: a structural equation modelling analysis based on survey data of 986 Chinese farmers. Food Control **80**, 143–150 (2017)

43. Damalas, C.A., Koutroubas, S.D.: Farmers' behaviour in pesticide use: a key concept for improving environmental safety. Curr. Opin. Environ. Sci. Health **4**, 30–37 (2018)

44. Farina, Y., Abdullah, M.P., Bibi, N., Khalik, W.M.A.W.M.: Determination of pesticide residues in leafy vegetables at parts per billion levels by a chemometric study using GC-ECD in cameron highlands, Malaysia. Food Chem. **224**, 186–192 (2017)

45. Dugger-Webster, A., LePrevost, C.E.: Following pesticide labels: a continued journey toward user comprehension and safe use. Curr. Opin. Environ. Sci. Health **4**, 19–26 (2018)

46. Hernandez, J., Mortimer, M., Patelli, E., Liu, S., Drummond, C., Kehr, E., Calabrese, N., Iannacone, R., Kacprzyk, J., Alemany, M., Gardner, D.: RUC-APS: enhancing and implementing knowledge based ICT solutions within high risk and uncertain conditions for agriculture production systems. In: 11th International Conference on Industrial Engineering and Industrial Management, 2017. Valencia, Spain (2017)

47. Saeed, M.F., Shaheen, M., Ahmad, I., Zakir, A., Nadeem, M., Chishti, A.A., Shahid, M., Bakhsh, K., Damalas, C.A.: Pesticide exposure in the local community of Vehari district in Pakistan: an assessment of knowledge and residues in human blood. Sci. Total Environ. **587–588**, 137–144 (2017)

48. Shinbaum, S., Crandall, P.G., O'Bryan, C.A.: Evaluating your obligations for employee training according to the food safety modernization act. Food Control **60**, 12–17 (2016)

49. Boobis, A.R., Ossendorp, B.C., Banasiak, U., Hamey, P.Y., Sebestyen, I., Moretto, A.: Cumulative risk assessment of pesticide residues in food. Toxicol. Lett. **180**, 137–150 (2008)

The Recovery of the *Old* Limachino Tomato: History, Findings, Lessons, Challenges and Perspectives

J. P. Martinez, C. Jana, V. Muena, E. Salazar, J. J. Rico, N. Calabrese,
J. E. Hernández, S. Lutts, and R. Fuentes

Abstract In this chapter we present a brief summary of several aspects related to the recovery of the *Old* Limachino Tomato, a specie cultivated in the Limache Basin located in the Valparaíso Region, Chile. What is very interesting in relation to this specie is that it disappeared completely of the market 45 years ago! We highlight both technical and commercial aspects associated with its recovery. In particular,

J. P. Martinez (✉)
Centro Regional de Investigación La Cruz, Instituto de Investigaciones Agropecuarias, Chorrillos 86, La Cruz, Valparaíso, Chile
e-mail: jpmartinez@inia.cl

Centro Regional de Estudios en Alimentos y Salud, Av. Universidad 330, Curauma, Valparaíso, Chile

C. Jana
Centro Regional de Investigación Intihuasi, Instituto de Investigaciones Agropecuarias, Colina San Joaquín S/N°, La Serena, Chile
e-mail: cjana@inia.cl

V. Muena
Centro Regional de Investigación La Cruz, Instituto de Investigaciones Agropecuarias, Chorrillos 86, La Cruz, Valparaíso, Chile
e-mail: victoria.muena@inia.cl

E. Salazar
Centro Regional de Investigación La Platina, Instituto de Investigaciones Agropecuarias, Santa Rosa, 11610 Santiago, Chile
e-mail: esalazar@inia.cl

J. J. Rico
Federación Empresarial de Agroalimentación de la Comunidad Valenciana, Isabel la Católica, 6, 5ta planta, 46004 Valencia, España
e-mail: adjunto.direccion@fedacova.org

N. Calabrese
Institute of Food Production Sciences, CNR ISPA. Via Amendola, 122/O, 70126 Bari, BA, Italy
e-mail: nicola.calabrese@ispa.cnr.it

J. E. Hernández
Management School, University of Liverpool, Liverpool, UK
e-mail: J.E.Hernandez@liverpool.ac.uk

J. E. Hernández and J. Kacprzyk (eds.), *Agriculture Value Chain — Challenges and Trends in Academia and Industry*, Studies in Systems, Decision and Control 280, https://doi.org/10.1007/978-3-030-51047-3_7

the project behind this huge endeavor has returned to the table of the Chileans a
fruit with high healthy and functional qualities. In addition, its strong connection
with a specific territory makes it very desirable for consumers. However, issues
related to the protection of the Peasant Family Farming of the cultivation area and
the improvement of post-harvest quality are key to guarantee the sustainability of
the business. In particular, we claim that in a sustainable agriculture development
perspective, there is an urgent need to increase the tolerance of this species to salinity
stress in order to increase productivity and reduce water use and fertilizers. Local
rootstocks appear as a promising technology to overcome part of these challenges.
Trademarks and certificates of protection are also urgent.

Keywords Limachino tomato · Peasant family agriculture · Rootstocks · Value
chain

1 Introduction

In 2015, the Food and Agriculture Organization of the United Nations (FAO), founded
in 1945, launched the International Year of Family Farming in order to "stress the vast
potential family farmers have to eradicate hunger and preserve natural resources".
According to that organization, small-scale family farming is innately connected
to global food security. In addition, it preserves traditional foods and contributes
to a balanced diet as well as to the protection of global agricultural biodiversity
and to the sustainable use of natural resources. Finally, it represents an opportunity
to energize local economies, especially when combined with policies specifically
targeted toward social welfare and community wellbeing. Quantitatively, more than
90% of world agricultural exploits are managed by one person or family, producing
around 80% of the world's food and occupying between 70 and 80% of agricultural
lands.

In conjunction with this launch, in July of that year the Foundation for Agricultural
Innovation (FIA, in Spanish), associated with the Ministry of Agriculture of Chile,
opened a call for Peasant Family Farming—Valuing the Agrarian Heritage. The
Agriculture Research Institute (INIA, in Spanish) in partnership with the Federico
Santa Maria University (UTFSM, in Spanish), responded to this call, and won, with

S. Lutts
Groupe de Recherche en Physiologie végétale, Earth and Life Institute-Agronomy (ELI-A),
Université catholique de Louvain, 5 (Bte 7.07.13) Place Croix du Sud, 1348 Louvain-la-Neuve,
Belgium
e-mail: Stanley.Lutts@uclouvain.be

R. Fuentes (✉)
Departmento de Industrias, Universidad Técnica Federico Santa María, Avenida España, 1680
Valparaíso, Chile
e-mail: raul.fuentes@usm.cl

their project, "Local, health, and sensory value of the Limachino tomato[1] for peasant family farming in Marga-Marga province".[2] An important point to note is that neither Juan Pablo Martinez of INIA nor Raúl Fuentes (both of whom were advisors and leaders of the project) were completely aware at the time of the impact a modest project like this (in terms of funds) would have, ranging from the recovery of the Old Limachino tomato's role in the "once" teatime ritual in Chile to more complex— and sometimes disappointing—political impact. This make this story deserve to be attended.

To put it briefly, the project's goal was the rescue and definitive reappearance to Chilean dining tables of this former icon of the Limache watershed in the Valparaiso region: the Old Limachino Tomato.[3] INIA led the agro-technical side of the project while UTFSM led the commercial. In particular, INIA was in charge of recovering old limachino seeds from international and national seed banks and from fields inside the Limache Basin through the collecting of the seed material held in producer's hands. Besides, INIA was in charge of recovering the ancestral agronomic management of this type of tomato. UTFSM was in charge of the creation and implementation of a business model for future commerce associated with the fruit. To highlight the enormity of this challenge, our goal here was to recover and reintegrate to the market a product that had disappeared 45 years ago! All the literature on management, marketing and economic development was against us. We had neither national nor international referents or similar examples with which to begin our investigation. We had to innovate from practically zero.[4]

Although these difficulties, we learned early on that the business of the Old Limachino Tomato was not only about producing and selling a fruit with supe-rior health and sensory traits, but also about selling a product with a strong local identity associated with a rich multicultural, intangible heritage. Above all, it was about selling a product that was (and is) loved by the people. The fruit was part of the collective imagination of many people older than 45. In other words, we learned that the key to the commercial success of the Old Limachino Tomato was in hitting the perfect note of subjectivity and emotion with the consumers along with business rationality. Having learned this lesson, it was much simpler to create our model. We made radical changes to two dimensions of the tomato's traditional commercial model: in its value proposition and in the configuration of its value chain. We realized that that the business was not about selling Old Limachino Tomato as such, but it was

---

[1] It refers to a variety of local tomato cultivated massively in the Limache Basin (place located in the central zone of Chile which is formed by the communes of Limache and Olmué) between 1950 and 1981 which become the main tomato traded in Chile before the appearance in the market of the long-lived varieties.

[2] For more details, see https://www.fia.cl/convocatorias-fia/ver-convocatoria/afc/.

[3] The adjective *old* was introduced to differentiate the genuine Limachino tomato from other long-lived varieties of tomatoes cultived in the Limache Basin.

[4] To the best of our knowledge, we knew some few cases of reintegration to the market of *new versions* of old products related to the automotive industry (new Austin Mini, new Fiat-600) and the music industry (new cassette; new long-play). In our case, our goal was the reintegration to the Chilean market of an old agro-product without the slightest modification.

about giving value to a rich historical-cultural patrimony of Chile. Having learned this, we decided, in a first stage, to sell the product directly, that is, without using intermediaries. The strategy was the following: once the product was well-valued by the consumers, the intermediaries would come to us and we would set the sale price. We took the risk and... we won.[5] The full implementation of the business model among the small local producers of the Old Limachino Tomato is still in development.

The project ended formally in January of 2017. It was declared an *emblematic* project by the then-authorities of the FIA. It had precisely fulfilled all its specific objectives. Today we know, in genetic terms, what is and what is not the Old Limachino Tomato. The geographic cultivation zone has been established. Over the course of the two-year of the project, 55–60,000 kg of fruit were sold at an average price never imagined by the producers, which significantly reduced the market uncertainty inherent in any innovation. Today, a long-term contract with an important international supermarket chain is currently working in order to initiate and consolidate the expansion of the product's commercialization to other regions in Chile, particularly to the capital city Santiago. In practice, the business model's proposed strategies worked.

Nevertheless, the Old Limachino Tomato's definitive return is not at all guaranteed. There are multiple and varied challenges to overcome so that the fruit doesn't return again to the tomb where it lay for almost 35 years. Among other challenges, the product's post-harvest durability must be improved. To this end, we have currently designed a strategy based on the use of rootstocks created from interspecific crosses between *S. chilense* and the salt-tolerant local tomato cultivar from Northern of Chile. According to our preliminary results, the use of this kind of rootstocks would increase significantly the lifetime of the product without altering its flavor and healthy attributes. On the other hand, advancements in biological managements of production are also needed and demanded by consumers. Creating channels to reach and conquer a younger public is another task to take on. Creating an appropriate and complete Value Chain is another one. But without a doubt, the most important challenge to overcome has to do with the design of a management model that achieves the highest level of connectedness, coordination, and commitment among the small producing business owners. Due to the complexity of the problems above mentioned, they are being tackled in collaboration with internationals institutions such as FEDACOVA (member of the European Cluster Collaboration Platform), Institute of Food Production Sciences (CNR-ISPA), the Liverpool University (from UK) and the Université catholique de Louvain, (from Belgium).

Meanwhile, in the short and medium terms, the Old Limachino Tomato still requires the cover and protection of institutions such as INIA and UTFSM. So far, both institutions, in the name of the local producers, obtained a trademark associated to the fruit and are gaining strength and momentum to obtain a certificate of protected

---

[5] At this point, it is worth mentioning that the attitude of the producers was rather of disbelief towards us because they had never in their lives been able to set the selling price of their products. Even worst, the price of tomatoes has always been set in the markets of Santiago, the capital of Chile.

designation of origin (PDO) for the fruit. The achievement of the latter would be a historic milestone for the Valparaiso region and for the country, as it would be the first vegetable to obtain that distinction. These efforts are being coordinated by the respective offices of technology transfer of both Institutions.

Taking everything into perspective, the business of the fresh Old Limachino Tomato is valued at 100–150 million Chilean pesos annually. If processed products (sauce, marmalade, preserves, and so on) are added to that figure, the valuation could reach one million dollars a year in a five-tear horizon. These are meaningful sums in family agriculture. However, the real underlying business of the Old Limachino Tomato will be in the returns on the intangible assets the project generated and continues to generate in the Limache watershed. Particularly, and in the words of its local government, the Limache commune and its surroundings "came back to life" with the recovery of the fruit.

In what follows, we reveal some very technical outcomes of the recovery project of the ancient limachino tomato, as well as the challenges to be overcome to consolidate its definitive return to the table of the Chileans.

2 Brief Characterization of the Old Limachino Tomato

The process of characterization of the Old Limachino Tomato started with the evaluation of productive, phenological, and morphological parameters in 13 tomato accessions gathered from international and national seed banks and from fields inside the Limache Basin through the collecting of the seed material held in producer's hands. From this evaluation, the formation of several groups were observed. Ten of the initial 13 accessions were discarded as potential limachino tomato types. One of the remaining accessions was identified as the Italian genotype *Cuore di Bue (Corazón de Buey, in spaninh)*, genotype collected locally in 2015 and codified with the acronym FIALIM3. See Fig. 1. The fruit related to this accession is named "Italian-type" limachino tomato by local farmers The others two remaining accessions, one of which was collected by INIA in 1960 and the other locally (from local farmers) in 2015, were identified as the Italian genotype *Costoluto Genovese*. These accessions were codified with the acronyms SLY74 and FIALIM4 and recognized as limachino tomato of the "French-type" by local farmers.[6] See Fig. 2.

Speaking shortly, once the study of the morphological and molecular characterisation was finished, the *Cuore di Bue* and the two *Costoluto Genovese* accessions are recognised within the family of the Old Limachino Tomato. However, from an ancestral management point of view on the part of the farmers, said accessions are

---

[6]Strictly speaking, both species were brought by Italian immigrants who arrived in the Limache Basin at the beginning of the twentieth century. The name "French-type" is due to the fact that this variety is very similar to the Marmande variety, a French variety that arrived at the Limache Basin during the 70s and that was massively cultivated by local producers.

Fig. 1 Tomato Cuore di Bue ("Corazón de Buey")

Fig. 2 Tomato Costoluto
Genovese

indeed recognised as such. From this results, the three species were grouped separately from commercial varieties and traditional varieties of European or American origin, and are currently being re-cultivated by the farmers of the Limache Basin.

On the side of agronomic management, the productivity of the Old Limachino Tomato, like any tomato variety, depends on several factors or determinants such as density, conduction, irrigation, nutrition, disease and pest control, soil, climate, among others. However, the Old Limachino Tomato's agronomic management has also some specific features that are important to be highlighted. According to the experimental data and data compiled from farmers, it was determined that the planting density of the Old Limachino Tomato is different for each type of accession: the "French-type" (FTOLT) and "Italian-type" (ITOLT). This, depending on the vigor of the respective plant. According to this criterion, our tests suggest that the planting density of the Italian-type accession should be lower than of the French-type accession due to its greater (relative) vigor. On the other hand, it was found, in practice that the adequate conduction in both accessions should be at one axis. This condition facilitates the agronomic managements made by producers, either managed outdoors or under greenhouse conditions. The outdoor conduction is carried out in an old-fashioned way, using in tutelage of "colihue" (*Chusquea culeou* in scientific terms) or wooden poles, and inside the greenhouse through the hanging of the plant through a linen cloth. Our results showed that a better functional and sensory quality of fruit is obtained at harvest in both types of tomatoes when the plant is allowed to grow only up to the fourth fruit cluster. However, in the case of the French-type (*Costoluto Genovese*) accessions, it is highly recommended to take out the big flower (named "florón"), because this organ takes away strength from the other flowers of the bunch. In general, it produces in the first cluster of this type of tomato. This management increases fruit size and the productivity. In relation to pests and diseases that affect the cultivation of the Old Limachino Tomato, they must be well managed to minimize the harmful effects they have on plants, always trying to avoid environment's contamination with chemical fertilizers, insecticides and chemical fungicides available in the market and minimizing the costs of fertilization and control so as not to affect crop productivity. The best way to control is using the Integrated Management of Pests and Diseases (Spanish acronym: MIPE), protocol oriented to reduce chemical products and increasing biological products. At this point, it is important to highlight that the ITOLT was the most unstable accession to the yield when subjected to conditions of organic management (compost) in comparison to the Long-Life Tomato and the French-type tomato. From this, it can be concluded that the French-type tomato has a higher rusticity than the Italian-type. This difference in rusticity and its effect on productivity was notoriously perceived.

In relation to the agronomic characterization, experiments carried out with the farmers of old limachino tomato in the Limache Basin show that the productivity up to the fourth bunch of the French-type limachino tomato and the Italian-type limachino tomato is around 2.0 and 1.6 kg per m^2 with 13 and 15 fruits per m^2 respectively, under optimal conditions of greenhouse cultivation. It is interesting to note that, at a productive level, the French-type limachino tomato (*Costoluto Genovese*) presents a similar productivity with respect to more commercial varieties like the Long-Lived

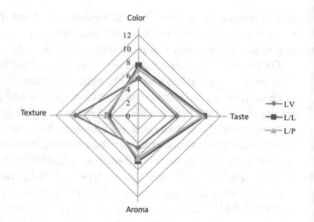

Fig. 3 Sensorial fruit quality (color, taste, aroma and texture) among non-grafted (long life variety, LV), selft-grafted (L/L) and grafted (L/P) Old Limachino Tomato plants

tomatoes. With respect to the quality of the fruit (functional and sensorial), it is well known that the importance of tomatoes as a healthy food is mainly explained by its different health-promoting compounds [21], such as lycopene, a-carotene, phenols, or Vitamin C. Currently research carrying out at INIA laboratories observes that the Old Limachino Tomato has a much higher antioxidant capacity than long-lived (LV) tomatoes on 50% [13] which, join with its unique and delicious taste and intensive red color and aroma (See Fig. 3), makes the fruit a very attractive and desirable product. However, the fruit presents a bad texture (See Fig. 3), short harvest period (one month) and low durability (5–7 days) compared to the Long-lived tomato. There-fore, it is necessary to surmounting gradually these problems in order to ensure its commercial sustainability. In line with this vision, INIA is currently studying the effect of the salt-tolerant rootstock on fruit quality at harvest and their healthy properties and fruit quality attributes after harvest in Old Limachino Tomato under saline stress, considering the importance of saline soils, environmental tendencies and global change around the world and Chile. In a sustainable agriculture devel-opment perspective, there is an urgent need to increase the tolerance of this species to salinity stress in order to reduce water use and fertilizers for agricultural produc-tion. Most of the local varieties cultivated since five decades ago were selected for yield potential under optimal conditions and they are thus lacking the key properties allowing the plant to cope with stress conditions. A promising alternative is to analyze the properties of old local varieties or closely-related wild relatives that could be used for breeding purposes. Centers of origin of tomato are located in Chile and it may thus be expected that plant material exhibiting a high level of tolerance to environ-mental and a high water use efficiency is still available in these areas. As far as local tomato varieties like the Old Limachino Tomato is concerned, grafting procedures can improve tomato crop productivity under various environmental conditions. The selection of appropriate rootstocks that allow the plant to cope with an abiotic stress such as salinity appears then as a priority. Halophyte plant species that could be crossed with the cultivated *S. lycopersicum* constitute a promising material for this purpose. INIA has access to a large germplasm collection of wild species and has already initiated studies focusing on Solanum chilense, [8–10, 14, 15] a halophyte wild tomato spontaneously growing in the North part of Chile, in the salt desert of Atacama, one of the driest areas in the world.

3   Prolonging Old Limachino Tomato's Lifetime by Creating and Using New Types of Rootstocks

As said, one urgent challenge to be surmounted is related to the short durability of the fruit once harvested. More specifically, the lifetime of the fruit ranges among 5–7 days under optimal cares during transport to the place of sale and under a delicate treatment from the part of the final consumer. This weakness makes no possible the commercial expansion of the product to more populated and distant markets. Based on international experiences, INIA is currently carrying out a couple of research projects aimed to introduce the use of new rootstocks created from interspecific crosses between *S. chilense* and the salt-tolerant local tomato cultivar from Northern of Chile. Three main effects are expected once the rootstock is grafted in the plant: gains in productivity, resistance to salinity and improvements in the quality of the fruit. In particular, consumers' preferences for fruits and vegetables with high concentrations of healthy compounds is steadily increasing and, therefore, it is important to test the effect of grafting on productivity and fruit quality under salinity conditions present in the growing zone of the Old Limachino Tomato. In line with this objective, it is well known from the works of Albacete et al. [1, 2] and He et al. [11] that grafting procedures can improve tomato crop productivity and postharvest quality under different environmental conditions, such as salinity.[7] On the other hand, at the optimal growing conditions, Barrett et al. [4] observed that fruit composition and sensory attributes of tomatoes are affected by grafting. They didn't reported changes in vitamin C, soluble solids, pH or titratable acidity among nongrafted, selft-grafted and grafted plants, but he detected differences in sensory attributes such as taste, color, aroma (See Fig. 3), appearance, acceptability, and flavor. Savvas et al. [20] observed that yield and fruit quality were negatively affected in the nongrafted and the self-grafted plants under saline stress conditions, increasing the total soluble solids and ascorbic acid content, while grafted plants the yield of two rootstocks was higher and the fruit quality was not affected on any quality characteristics. On the other hand, Martínez-Rodriguez et al. [17] and Di Gioia et al. [6] investigated mechanisms involved in the effects of grafting on yield performance in tomatoes

---

[7]From the genetic perspective, it has been demonstrated that the difficulties to enrich elite lines with genes from wild species to confer tolerance are linked with the large number of genes involved, most of them with small effect compared to the environment, and high costs of recovering the genetic background of the receptor cultivar [5]. A practical alternative for breeders seems to be the introduction of genes associated with salt tolerance to rootstocks, thereby converting a sensitive cultivar into a tolerant one, maintaining all the highly valuable characteristics which current cultivars possess. New traits are rarely introduced from wild germplasm, requiring many generations to remove the deleterious genes that go along with the introduced genes due to linkage drag.

reporting that the salt tolerance conferred by some graft combinations is related to the ability of the rootstock to reduce the uptake and transport rates of saline ions to the shoot. This salt exclusion mechanism of rootstock might mitigate salt stress [6, 17], which could be a physiological mechanism of salt tolerance in rootstock. Related to the Old Limachino Tomato, we have observed in our hat K^+ extraction in "Old Limachino Tomato" grafted plants were higher than in self-grafted one optimal greenhouse conditions. These antecedents suggest that the use of the salt tolerant rootstock may improve the yield and fruit quality at pre and postharvest under salt stress conditions. Taking into account this hypothesis, INIA is currently starting a research project aimed at revealing mineral regulation (K^+ and Na^+) at the root level using "Old Limachino Tomato" grafting as a model system with emphasis on pre and postharvest fruit quality parameters using a rootstock from an interspecific cross from *S. lycopersicum* and *S. chilense*. So far, our preliminary results suggest that "Old Limachino Tomato" might be characterized as a tomato with a low saline tolerance which reduces strongly its productivity. This, based on compared experiments made with compost with high electrical conductivity and with conventional fertilization management [12].

Since current commercially available rootstocks still exhibit a lack of attributes not allowing the plant to avoid the negative impact of salinity in the Northern and Central zones of Chile, grafting procedure has been recommended as a promising strategy to improve salt tolerance in this local variety. However, the use now of appropriate rootstocks/scion combinations may offer the advantage of combining rustic and vigorous root systems and high-producing shoots, allowing farmers to maximize benefits in non-optimal conditions and to reduce inputs in terms of water and fertilizer. It may thus be hypothesized that a specific selection must be performed to select salt-resistant rootstocks to be used for tomato production in salt-affected areas. In fact, cultivated tomato has numerous wild relatives able to cope with high salinity levels and which could thus be classified as halophyte. Some promising plant species have been therefore exhaustively studied and numerous data are available on *S. pennellii* (Tapia et al. [22], *S. pimpinellifolium* [7], *S. cheesmanii* [2] or *S. galapagense* [18]. For some of these species, interspecific crosses with cultivated *S. lycopersicum* have been successfully performed and hybrids have been characterized [3], although important work is still necessary for the identification of salt-resistant material exhibiting suitable yield capacities. The Chilean halophyte species *S. chilense* surprisingly has received only minor attention from the scientific community. This species is native to the Atacama Desert and is able to grow under salt levels similar to those encountered in sea water. This fascinating ability to manage with a high salt concentration was until now underexploited, mainly because of the allogamous nature of this wild species and the small size of produced fruit, which are undesirable traits to integrate in breeding schemes. Studies carried out by Martinez et al. [15], Tapia et al. [22] and Gharbie et al. [8–10] have shown that wild relative tomato *S. chilense* is tolerant to salinity. For this reason, INIA performed some interspecific crosses between *S. chilense* and the salt-tolerant local cultivar from Northern of Chile. The obtained interspecific crosses exhibited a high relative growth rate and root-to-shoot ratio during preliminary trial. Accordingly, the main hypothesis under revision is that this material may be used as a salt-tolerant rootstock allowing yield and fruit quality (pre and postharvest) improvements in the salt-sensitive local cultivar ("Old Limanchino Tomato"). Grafting with selected rootstocks issued from interspecific crosses between cultivated *S. lycopersicum* and wild-relative halophyte *S. chilense* has been proposed as an agronomical alternative to improve the salt tolerance of "Old Limachino Tomato".

On the quality side, a few studies on a relatively limited number of rootstocks have been published to report the effect of grafting on tomato fruit quality and its relation to tolerance to saline conditions. According to Savvas et al. [20], using rootstock tolerant to salinity and regulating the ratio of Na^+ to K^+ (lower) can be effective tools to enhance the tolerance of the shoot part and fruit quality. Another report by He et al. [11]) indicates that the alleviation of salinity induced growth inhibition in tomato plants grafted on a Chinese commercial rootstock was related to an enhancement of antioxidant capacity and concomitant improvements in photosynthesis. Preliminary results show that the local rootstock has an effect on the fruit quality (color, taste, aroma and texture) among nongrafted, selft-grafted and grafted plants. The lack of information on the effects of rootstocks on fruit´s quality cultivated under high salinity conditions offers a great room for research.

4   Market, Business Model and Value Chain for the Old Limachino Tomato

4.1   Market and Business Model

As well known, an efficient way to start a particular (and private) business is to insert it within a certain conceptual framework designed for making businesses, making there the necessary adjustments that account for the specific characteristics that the business to raise has and/or requires. In this way, the business of the Old Limachino Tomato proposed at its moment had the first thing that every business and the organization that supports it must have: a logic. In other words, it is necessary to define a Business Plan based on a Business Model. What is sought with the design of a Business Model? Something very simple and complex at the same time: describe the logic of how an organization creates, delivers and captures value in some economic, social, cultural and political context. What is sought with the design of a Business Plan? Define the most important actions that must be performed to create, deliver and capture the said value. In our case, we used the well-known Canvas Business Model (CBM), hereinafter), model based on the work of Osterwalder [19].

In short, the structure of the CBM seeks to identify the logics of interaction between the different parts that make up a business and the organization that supports it. For this, axes (or pillars) and sub-axes are defined. The four axes are: Offer (or Value Proposal), Infrastructure, Customers and Financial Viability. The respective sub-axes are: (a) key partners, key activities and resources; (b) relations with customers, distribution channels and customer segmentation. Are the axes of the model hierarchically ordered? Yes. The main axis (or main pillar) of the model is the Offer or Value Proposal. In simple and somewhat reductionist words, this axis seeks to clearly define what is being sold or offered to the market. At a higher conceptual level, this axis seeks to clearly identify the sources of business value creation, from the most tangible (relatively easier to quantify monetarily) to the most intangible (those most difficult to quantify monetarily).[8] Because a detailed explanation of all the contents (axes and sub-axes) of the business model for the Old Limachino tomato is beyond the scope of this chapter, we just characterized here -in detail- its main component: the Proposal Value. That said, the sources of value creation identified and associated with the Old Limachino Tomato and its externalities to date are the following:

- Genuine product, unique in the world. Tomato Limachino will be produced under cultivation patterns rooted in the historical-cultural heritage of the producers of the Limache Basin. Currently, its genetic characterization (fingerprint) is being terminated to avoid plagiarism.
- Product with historical heritage. A product with a strong historical story will be offered to the market. Currently, this story is being reconstructed in conjunction with the former farmers of Tomato Limachino in order to restore the ancestral agronomic managements that give the tomato a value of belonging to its territory.
- Product with specific quality. The Old Limachino Tomato is being differentiated from any other tomato through the characterization of its healthy attributes (antioxidant capacity, low sugar content, titratable acidity and carotenoids, among others). To date, statistical data are already available that indicate the superiority of the sensory properties of tomato limachino (flavor, aroma and color) of those related to the most commercial long-lived tomatoes).
- Product with territorial value. It refers to the fact that the cultivation area will be strictly limited to the current Limache Basin, which encompasses the current neighboring communes of Limache and Olmué.
- Product with a better healthy quality. It refers to the innovative improvements in pre and post-harvest under development will increase the current healthy attributes of the fruit.
- Progressively grown product with biological agronomic practices. It refers to the fact that producers are committed to completely and progressively eliminating the use of chemical fertilizers and other non-biological practices.
- Product associated to the recovery of an icon referent of the Limache watershed in the Valparaiso region which generates an intangible asset to be managed jointly by the regional and local authorities to optimize its returns.
- Product with the potential to creating a Gastronomic Route and related undertakings in Limache and Olmué and other communes of the central zone of Chile.

[8] All this knowledge may sound too evident for people who manage medium or large firms. However, every single word of all of this information was completely unknown for the twelve-very-small producers engaged in the project.

INIA and the UTFSM succeeded in transmitting a great deal of this powerful message to the current twelve farmers engaged in the project. In addition, and more importantly, both institutions succeeded in transmitting the message to consumers throughout different channels. Armed with this information, the farmers themselves were able to get, after the bargaining process with one of the biggest Latin-American supermarkets, a selling price never imagined before. More concretely, the selling price set to this intermediary is today about eight times the current selling price of the long-lived tomato varieties that they also cultivate and sell. Therefore, the Financial Viability of the Old Limachino Tomato business is guaranteed and the small farmers engaged in it should focus their attention on other components of the Business plan. One of them is related to logistic aspects of the commercialization which can be tackled by designing properly the value chain for the fruit.

4.2 Value Chain of the Old Limachino Tomato

The value chain of the Old Limachino Tomato for fresh consumption is a simple system and is mainly associated with small-scale agriculture (Fig. 4). The value chain starts from the seed (Box 1: seeds), which considers the use of a genuine genetic material. With this material, the seedlings are produced (Box 2: nursery plant process), sometimes this process is carried out by the producers or the companies producing the seedlings. The waste in this process is mainly associated with: use of seed of poor quality, uncontrolled cultivation and environmental conditions. Box 3 (seedling transport), corresponds to the transportation of seedlings from the company producing seedlings to the farmer. The waste in this step occurs when transport conditions are not adequate. Box 4 (tomato production process + harvest) (Fig. 4) shows the processes associated with tomato production, such as agronomic

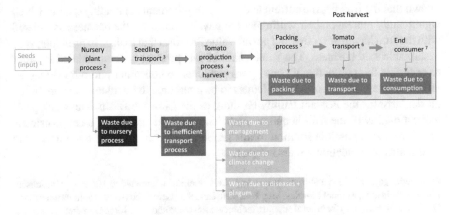

Fig. 4 Simplified schema of the *Old* Limachino Tomato value chain

management and crop conditions. In this phase, the wastes are principally associated with problems in the agronomic management, occurrence of pests and diseases, climate change, among others, including also the losses in the harvest. Box 5 (packing process), wastes are observed when the storage conditions are not adequate, and the fruit shows deterioration in its physiological state, as well as if the selection prior to storage was not appropriate, quality losses are observed (example: store fruit with presence of fungus, produces a spread of the disease in the storage process). The process of the fruit transporting corresponds to Box 6 (tomato transport), use of unsuitable transportation means, which presents wastes that are produced by not using an adequate packaging, uncontrolled environmental conditions of transportation means, and excessive transports. Finally, in Box 7 (end consumer), some wastes occur due to the fruit not consumed and to problems of condition and quality for the consumer [16].

5 Conclusions

Thanks to a joint work between the Agriculture Research Institute (INIA) and the Universidad Federico Santa María (UTFSM), the Tomato Limachino -fruit disappeared of the Chilean tables 45 years ago–is back today. This work involved both technical and commercial efforts. In technical terms, today's renamed *Old* Limachino Tomato (in its French and Italian-types) has been recovered and wholly characterized in its functional, healthy and sensorial capacities. It is known today that it has a higher antioxidant capacity and higher antioxidant content such as polyphenols than the best more commercial tomato varieties. It is also known that it has a rich flavour and aroma that make it appealing to consumers. In commercial terms, it was possible to identify the fruit with a rich historical and cultural heritage associated with the Limache Basin, a place located in the Region of Valparaíso, Chile. Besides, it is known that the fruit fits the current and trendy requirements of healthy quality which significantly increases their willingness to pay. In summary, the business associated with the fruit has an attractive and power Value Proposal that makes, in principle, very sustainable said business. The value chain of this product is being studied in order to optimize the productive process from producer to consumer and reduce wastes. However, there are important challenges to be overcome to guarantee sustainability, particularly for the Peasant Family Farming of the Limache Basin. The weak postharvest quality of the fruit is one of them. The creation and use of local rootstocks are viewed as a possible solution to this problem. The obtaining of trademarks and certificates of protection are also urgent.

Acknowledgement Juan Pablo Martínez and Raúl Fuentes acknowledge Helmuth Hinrichsen, Karina Pasalacqua, Manuel Morales, Meg Pedersen, Luis Salinas and the twelve small farmers of the Old Limachino Tomato for helpful support, technical assistance and love. Jorge Hernández acknowledge the contribution of the Project 691249, RUC-APS: Enhancing and implementing Knowledge based ICT solutions within high Risk and Uncertain Conditions for Agriculture Production Systems (www.ruc-aps.eu), funded by the European Union under their funding scheme H2020-MSCA-RISE-2015. Raúl Fuentes acknowledges the support of the General Direction of Research, Innovation, and Graduate Studies (DGIIP), the Office of Technology Transfer (OTTL), and the Department of Industrial Engineering of the Universidad Técnica Federico Santa María. This work was partly funded by the National Commission for Scientific & Technological Research (CONICYT) of the Ministry of Education of Chile through the FONDECYT project N° 1180958.

References

1. Albacete, A., Ghanem, M.E., Dodd, I.C. Pérez-Alfocea, F.: Principal component analysis of hormone profiling data suggests an important role for cytokinins in regulating leaf growth and senescence of salinised tomato. Plant Signal Behav **5**, 44–46 (2010)
2. Albacete, A., Martínez-Andújar, C., Ghanem, M.E., Acosta, M., Sánchez-Bravo, J., Asins, M.J., Cuartero, J., Lutts, S., Dodd, I.C., Pérez-Alfocea, F.: Rootstock-mediated changes in xylem ionic and hormonal status are correlated with delayed leaf senescence and increased leaf area and crop productivity in salinised tomato. Plant Cell Environ. **32**, 928–938 (2009)
3. Asins, M.J., Bolarín, M.C., Pérez-Alfocea, F., Estañ, M.T., Martínez-Andújar, C., Albacete, A., Villalta, I., Bernet, G.P., Dodd, I.C., Carbonell, E.A.: Genetic analysis of physiological components of salt tolerance conferred by *Solanum rootstocks*. What is the rootstock doing for the scion? Theory Appl. Genet. **121**, 105–115 (2010)
4. Barrett, C.E., Zhao, X., Sims, C.A., Brecht, J.K., Dreyer, E.Q., Gao, Z.: Fruit composition and sensory attributes of organic heirloom tomatoes as affected by grafting. Hort. Technol. **22**, 804–8095 (2012)
5. Cuartero, J., Bolarín, M.C., Asíns, M.J., Moreno, V.: Increasing salt tolerance in the tomato. J. Exp. Bot. **57**, 1045–1058 (2006)
6. Di Gioia, F., Signore, A., Serio, F., Santamaria, P.: Grafting improves tomato salinity tolerance through sodium partitioning within the shoot. Hort. Sci. **48**, 855–862 (2013)
7. Gálvez, F.J., Baghour, M., Hao, G., Cagnac, O., Rodríguez-Rosales, M.P., Venema, K.: Expression of LeNHX isoforms in response to salt stress in salt sensitive and salt tolerant tomato species. Plant Physiol. Biochem. **51**, 109–115 (2012)
8. Gharbi, E., Martínez, J.P., Benahmed, H., Fauconnier, M.L., Lutts, S., Quinet, M.: Salicylic acid differently impacts ethylene and polyamine synthesis in the glycophyte *Solanum lycopersicum* and the wild-related halophyte *Solanum chilense* exposed to mild salt stress. Physiol. Plant **158**, 152–167 (2016)
9. Gharbi, E., Martínez, J.P., Benahmed, H., Lepoint, G., Vanpee, B., Quinet, M., Lutts, S.: Inhibition of ethylene synthesis reduces salt-tolerance in tomato wild relative species *Solanum chilense*. J. Plant Phisiol. **210**, 24–37 (2017)
10. Gharbi, E., Martínez, J.P., Benahmed, H., Dailly, H., Quinet, M., Lutts, S.: The salicylic acid analog 2,6-dichloroisonicotinic acid has specific impact on the response of the halophyte plant species *Solanum chilense* to salinity. Plant Growth Regul. **82**, 517–525 (2017)
11. He, Y., Zhu, Z., Yang, J., Ni, X., Biao Zhu, B.: Grafting increases the salt tolerance of tomato by improvement of photosynthesis and enhancement of antioxidant enzymes activity. Environ. Exp. Bot. **66**, 270–278 (2009)
12. Jara, C.: Efecto de la fertilizacion (convencional y compost) sobre la productividad y calidad de Tomate Limachino Antiguo y larga vida bajo condiciones de invernadero en la cuenca de Limache, p. 70. Tesis de Magíster (MTA), Universidad Técnica Federico Santa María, Santiago, Chile (2016)
13. Martínez, J.P., Muena, V., Salinas, L., Freixas, A., Farías, K., Loyola, N., Fuentes, R.: Efecto del uso de portainjerto INIA sobre el sabor de fruto de Tomate Limachino (*Solanum lycopersicum*) en relación al tomate de larga vida, 66° Congreso Agronómico SACH y 13° SOCHIFRUT, 17 al 20 de noviembre. Valdivia, Chile (2015)

14. Martínez, J.P., Antúnez, A., Araya, H., Pertuzé, R., Fuentes, F., Lizana, X.C., Lutts, S.: Salt stress differently affect growth, water status and antioxidant enzyme activities in *Solanum lycopersicum* L. and its wild-relative *Solanum chilense* Dun. Aust J Bot **62**, 359–368 (2014). https://doi.org/10.1071/BT14102

15. Martínez, J.P., Antunez, A., Acosta, M.P.P.R., Palma, X., Fuentes, L., Ayala, A., Araya, H., Lutts, S.: Effects of saline water on water status, yield and fruit quality of wild (*Solanum chilense*) and domesticated (*Solanum lycopersicum* var.*cerasiforme*) tomatoes. Expl. Agric. **48**, 573–586 (2012)

16. Hernández, J., Mortimer, M., Patelli, E., Liu, S., Drummond, C., Kehr, E., Calabrese, N., Iannacone, R., Kacprzyk, J., Alemany, M., Garder, D.: RUC-APS: Enhancing and implementing knowledge based ICT solutions within high risk and uncertain conditions for agriculture production systems. In: 11th International Conference on Industrial Engineering and Industrial Management, Valencia, Spain (2017)

17. Martinez-Rodriguez, M.M., Estañ, M.T., Moyano, E., Garcia-Abellan, J.O., Flores, F.B., Campos, J.F., Al-Azzawi, M.J., Flowers, T.J., Bolariìn, M.C.: The effectiveness of grafting to improve salt tolerance in tomato when an 'excluder' genotype is used as scion. Environ. Exp. Bot. **63**, 392–401 (2008)

18. Montforte, A., Asíns, M.J., Carbonnell, E.A.: Salt tolerance in Lycopersicon species. VI. Genotype by salinity interaction in quantitative trait loci detection. Constitutive and response QTs. Theory Appl. Genet. **95**, 706–713 (1997)

19. Osterwalder, A.: The Business Model Ontology—a Proposition In A Design Science Approach. Ph.D. thesis, University of Lausanne (2004)

20. Savvas, D., Savva, A., Ntatsi, G., Ropokis, A., Karapanos, I., Krumbein, A., Olympios, C.: Effects of three commercial rootstocks on mineral nutrition, fruit yield, and quality of salinized tomato. J. Plant Nutr. Soil Sci. **174**, 154–162 (2011)

21. Seguel, I.: Mecanismo Nacional de Intercambio de Información sobre la aplicación del plan de acción mundial para la conservación y utilización sostenible de los recursos fitogenéticos para la alimentación y la agricultura. 1–54 p In: Informe nacional sobre el estado de los recursos fitogenéticos para la agricultura y la alimentación. Seguel I y Agüero T. (Eds.). INIA y FAO, Chile. https://www.pgrfa.org/gpa/chi (2008)

22. Tapia, G., Méndez, J., Inostroza, L.: Different combinations of morpho-physiological traits are responsible for tolerance to drought in wild tomatoes *Solanum chilense* and *Solanum peruvianum*. Plant Biol. **18**, 406–416 (2015)

Printed in the United States
by Baker & Taylor Publisher Services